Bioremediation for Environmental Sustainability

Edited by Wafaa M. Abd El-Rahim

Published in London, United Kingdom

Bioremediation for Environmental Sustainability
http://dx.doi.org/10.5772/intechopen.1005554
Edited by Wafaa M. Abd El-Rahim

Contributors
Amanda Vieira de Barros, Dalel Daâssi, Dirk van Zyl, Fabricio Motteran, Fatimah Almaghrabi, Felicia Omolara Afolabi, Harlina Ahmad, Hassan Moawad, Jamile Maria Pereira Bastos Lira de Vasconcelos, Kaline Catiely Campos Silva, Karolayne Silva Souza, Lina Xie, Lívia Caroline Alexandre de Araújo, Maria Betânia Melo de Oliveira, Matheus Victor Aragão Cavalcanti, Milena Roberta Freire da Silva, Mohamed Chamkha, Mouna Bouassida, Paul Musonge, Wafaa M. Abd El-Rahim

First published in London, United Kingdom, 2025 by IntechOpen
IntechOpen is the global imprint of INTECHOPEN LIMITED, registered in England and Wales, registration number: 11086078, 167-169 Great Portland Street, London, W1W 5PF, United Kingdom

For EU product safety concerns: IN TECH d.o.o., Prolaz Marije Krucifikse Kozulić 3, 51000 Rijeka, Croatia, info@intechopen.com or visit our website at intechopen.com.

British Library Cataloguing-in-Publication Data
A catalogue record for this book is available from the British Library

Bioremediation for Environmental Sustainability
Edited by Wafaa M. Abd El-Rahim
p. cm.

This title is part of the Environmental Sciences Book Series, Volume 29
Topic: Environmental Resilience and Management
Series Editor: J. Kevin Summers
Topic Editor: Jose Navarro-Pedreño

Print ISBN 978-1-83634-290-8
Online ISBN 978-1-83634-289-2
eBook (PDF) ISBN 978-1-83634-291-5
ISSN 2754-6713

IntechOpen Book Series

Environmental Sciences

Volume 29

Aims and Scope of the Series

Scientists have long researched to understand the environment and man's place in it. The search for this knowledge grows in importance as rapid increases in population and economic development intensify humans' stresses on ecosystems. Fortunately, rapid increases in multiple scientific areas are advancing our understanding of environmental sciences. Breakthroughs in computing, molecular biology, ecology, and sustainability science are enhancing our ability to utilize environmental sciences to address real-world problems.

The four topics of this book series - Pollution; Environmental Resilience and Management; Ecosystems and Biodiversity; and Water Science - will address important areas of advancement in the environmental sciences. They will represent an excellent initial grouping of published works on these critical topics.

Meet the Series Editor

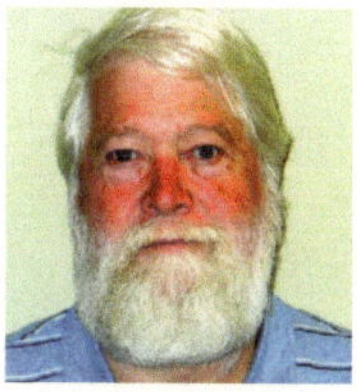

J. Kevin Summers is a Senior Research Ecologist at the Environmental Protection Agency's (EPA) Gulf Ecosystem Measurement and Modeling Division. He is currently working with colleagues in the Sustainable and Healthy Communities Program to develop an index of community resilience to natural hazards, an index of human well-being that can be linked to changes in the ecosystem, social and economic services, and a community sustainability tool for communities with populations under 40,000. He leads research efforts for indicator and indices development. Dr. Summers is a systems ecologist and began his career at the EPA in 1989 and has worked in various programs and capacities. This includes leading the National Coastal Assessment in collaboration with the Office of Water which culminated in the award-winning National Coastal Condition Report series (four volumes between 2001 and 2012), and which integrates water quality, sediment quality, habitat, and biological data to assess the ecosystem condition of the United States estuaries. He was acting National Program Director for Ecology for the EPA between 2004 and 2006. He has authored approximately 150 peer-reviewed journal articles, book chapters, and reports and has received many awards for technical accomplishments from the EPA and from outside of the agency. Dr. Summers holds a BA in Zoology and Psychology, an MA in Ecology, and Ph.D. in Systems Ecology/Biology.

Meet the Volume Editor

Prof. Wafaa Mohamed Abdel-Rahim has been a Professor of Environmental Microbiology at the National Research Centre (NRC) in Egypt since 2011. She previously served as the Head of the Agricultural Microbiology Department for six years and is currently the Institute of Biological and Agricultural Research Dean. She has led numerous national and international research projects, securing funding for 22 projects, including 14 as the principal investigator (PI). Her outstanding scientific contributions have earned her nine prestigious awards, including the State Encouragement Award and two international honors from the Korea Exhibition of Innovation. Prof. Wafaa is an active member of several national committees, such as the Academy of Scientific Research and Technology, the Microbiology and Soil Committees, and environmental and water councils. She holds three patents from the Academy of Scientific Research and Technology in Egypt. She has trained many students from various Egyptian universities since 2010. Additionally, she serves as the Chair of the Promotions Committee for Biochemistry and Biological Sciences under the Ministry of Scientific Research and Technology. She has participated in numerous international scientific missions, representing Egypt in global conferences and fostering research collaborations with Arab and foreign countries. Through these engagements, she has strengthened academic and scientific partnerships, contributing to the advancement of environmental microbiology on an international scale.

Contents

Preface

Bioremediation plays a vital role in achieving environmental sustainability by utilizing biological agents such as plants, microorganisms, and fungi to detoxify contaminated environments. The book Bioremediation for Environmental Sustainability explores various innovative approaches that enhance the efficiency of bioremediation processes, offering sustainable solutions to soil and water contamination. This compilation of research provides insights into phytoremediation, microbial remediation, and technological advancements in the field, making it a valuable resource for researchers, practitioners, and policymakers.

Chapter Summaries

1. *Phytoremediation: The Green Solution* – Written by Harlina Ahmad, this chapter highlights the role of plants in absorbing and neutralizing environmental pollutants. It discusses the mechanisms of phytoremediation, its effectiveness, and its applications in cleaning up contaminated soil and water.

2. *Application of Fruit Wastes in the Bioremediation of Heavy Metals: Mechanism, Challenges and Prospects* – This chapter explores the potential of fruit-derived compounds in removing heavy metals from polluted environments. It presents scientific evidence on how organic components from fruits can interact with toxic elements, offering an eco-friendly and cost-effective alternative for remediation.

3. *Geochemical Processes Controlling Metals (Pb, Zn) Mobility during Amended Phytostabilization on Sulfidic Mine Tailings* – Lina Xie and Dirk van Zyl examine the geochemical processes in the rhizosphere of sulfidic mine tailings during phytostabilization, revealing that plant activity lowers pH and increases metal release, while amendment additions neutralize acidity through three geochemical processes.

4. *Innovations in Herbicide Bioremediation: Green Solutions for Soil Contamination* – Karolayne Silva Souza et al. build on the concepts introduced in previous chapters, focusing on innovative strategies for degrading herbicides in contaminated soil. The chapter discusses novel biotechnological approaches that combine phytoremediation, microbial degradation, and mechanized support to enhance soil restoration.

5. *Mycoremediation: An Innovative and Sustainable Approach* – This chapter by Dalel Daâssi et al. introduces the concept of using fungi for bioremediation and explores the mechanisms by which fungi break down pollutants, emphasizing their role in degrading hydrocarbons, heavy metals, and other toxic substances in the environment. Further expanding on fungal bioremediation, the chapter discusses different fungal species and their efficiency in removing environmental contaminants. This chapter highlights real-world applications of mycoremediation in restoring polluted ecosystems.

6. *The Role of Bioremediation in Achieving Environmental Sustainability* – Wafaa M. Abd El-Rahim and Hassan Moawad conclude the book with a comprehensive discussion on the overall impact of bioremediation in achieving sustainability. This chapter synthesizes key insights from previous chapters and highlights future perspectives on integrating bioremediation with global environmental policies and sustainability goals.

Wafaa M. Abd El-Rahim
Agricultural Microbiology Department,
Agricultural and Biological Research Institute,
National Research Centre (NRC),
Cairo, Egypt

Acknowledgements

We extend our heartfelt gratitude to all contributors who played a vital role in developing this book. Special thanks to the editors for their dedication to refining the content, the authors for their invaluable research and insights, and Intechopen for their efforts in bringing this work to fruition. Their collective contributions have made this book a valuable resource in the field of bioremediation and environmental sustainability.

Chapter 1

Phytoremediation: The Green Solution

Harlina Ahmad

Abstract

Phytoremediation is an eco-friendly, economical method that uses plants to remediate pollutants in soil, water, and air. This chapter examines the fundamental processes of phytoremediation, including phytoextraction, phytostabilization, phytodegradation, phytovolatilization, and rhizofiltration, that allow plants to absorb, degrade, stabilize, or filter contaminants such as heavy metals and organic pollutants. The work explores the selection of suitable plant species, emphasizing hyperaccumulators, and provides examples of cases that exemplify effective practical applications. This chapter analyzes current biotechnological breakthroughs that have improved the efficacy of phytoremediation, particularly via the use of genetically modified plants. Furthermore, it highlights the ecological and economic advantages of this green remediation method, while outlining prospective research opportunities to enhance its efficacy and face complex pollution issues. The chapter discusses phytoremediation's significant role in sustainable environmental management and pollution mitigation through thorough exploration.

Keywords: phytoremediation, environmental restoration, pollutant detoxification, sustainable remediation, plant-based clean-up

1. Introduction

Phytoremediation is an innovative, plant-based strategy for addressing environmental pollutants, including heavy metals, organic contaminants, and radionuclides in soil, water, and air. Derived from the Greek word "phyto" (plant) and the Latin "remedium" (treatment), this method offers a sustainable, eco-friendly alternative to traditional remediation techniques. Its appeal lies in its cost-effectiveness, environmental sustainability, and ability to restore ecosystems with minimal disruption. By harnessing the natural processes of specific plants, phytoremediation provides an efficient, low-impact solution for contaminant removal or stabilization.

Phytoremediation improves polluted environments while enhancing ecosystem health and resilience by preserving biodiversity and supporting recovery from environmental stresses. For instance, a study conducted in the Lake Hawassa sub-basin showed that active restoration efforts increased biodiversity and soil moisture, significantly contributing to ecosystem resilience in water-limited areas [1]. By promoting species diversity and restoring critical ecosystem functions, phytoremediation enables ecosystems to recover from pollution and become more resilient to future challenges. It plays a crucial role in ecological restoration by fostering biological stability and resilience, aiding ecosystems in adapting to anthropogenic disturbances [2].

Phytoremediation encompasses a range of processes that vary depending on the specific pollutant and environmental medium (soil, water, or air) involved. Plants can absorb, stabilize, decompose, or volatilize contaminants, making this method effective for addressing pollutants such as heavy metals, hydrocarbons, and agricultural chemicals. Due to its versatility, phytoremediation can be applied across various settings, including industrial sites, agricultural fields, and urban areas. It is widely recognized as an effective environmental management tool, supporting the natural restoration of degraded ecosystems while maintaining biodiversity [3].

Phytoremediation plays a vital role in environmental restoration by offering a sustainable alternative to conventional remediation methods, which are often costly and environmentally harmful [4]. Traditional techniques, such as soil extraction, landfill disposal, and chemical treatments, require significant financial and energy resources and can have detrimental effects on the environment. In contrast, phytoremediation harnesses natural processes to restore environmental health without causing additional harm. By utilizing plants to stabilize or remove contaminants, this approach effectively eliminates pollutants from contaminated areas while enhancing ecosystem resilience and recovery [5].

Phytoremediation significantly improves soil health, preserves biodiversity, and contributes to carbon sequestration. By enhancing soil structure and increasing microbial activity, it supports soil fertility and ecosystem functionality. The introduction of carefully selected plant species into polluted areas facilitates the restoration of vegetation cover, providing habitats for organisms and promoting ecosystem recovery. Additionally, as plants grow and absorb pollutants, they simultaneously capture carbon dioxide from the atmosphere, aiding in climate change mitigation. Phytoremediation is often more cost-effective than mechanical or chemical methods, making it an ideal solution for large-scale projects or economically disadvantaged regions. With its potential for long-term use and bioenergy production, phytoremediation offers a promising approach to pollution reduction and environmental sustainability [6].

This chapter provides an in-depth exploration of the key mechanisms of phytoremediation, including phytoextraction, phytostabilization, phytodegradation, phytovolatilization, and rhizofiltration. It emphasizes the importance of selecting appropriate plant species tailored to specific contaminants and environmental conditions. Additionally, it examines advancements in biotechnology, such as genetic engineering and microbial-assisted technologies, that enhance plant capacities for contaminant removal. The ecological and economic advantages of phytoremediation are also discussed, highlighting its potential as a sustainable environmental solution.

2. Mechanisms of phytoremediation

Phytoremediation operates through a variety of biological and chemical mechanisms that enable plants to interact with environmental contaminants. These mechanisms exploit the natural processes of plants to stabilize, degrade, or remove pollutants from soil, water, or air. The primary mechanisms of phytoremediation include phytoextraction, phytostabilization, phytodegradation, phytovolatilization, and rhizofiltration. These processes rely on the plant's inherent metabolic systems,

DOI: http://dx.doi.org/10.5772/intechopen.1007671

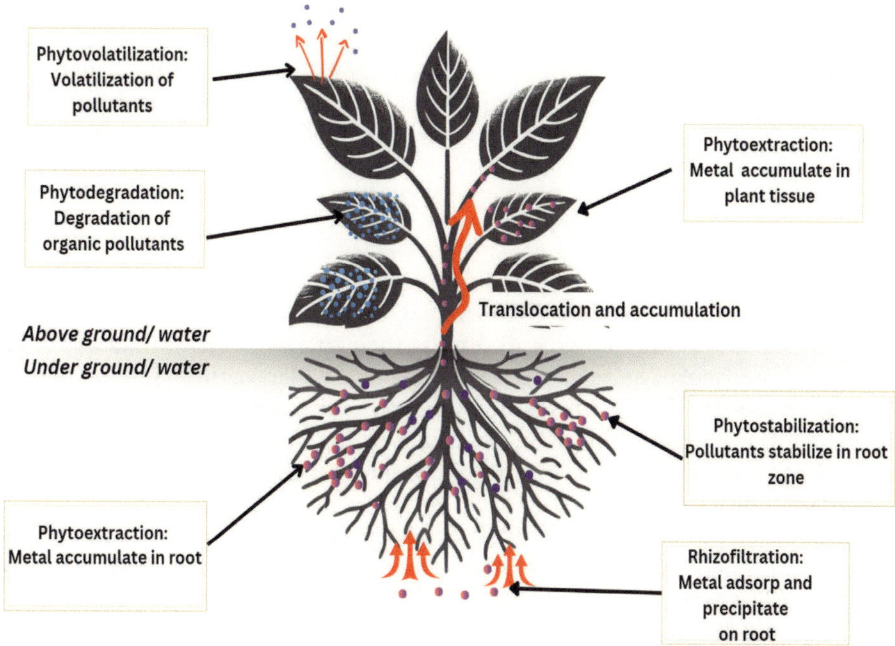

Figure 1.
Plant interaction with soil or water contaminants for each mechanism.

such as root uptake, microbial interactions within the rhizosphere, and enzymatic degradation. **Figure 1** illustrates how plants interact with soil and water contaminants through these different mechanisms.

2.1 Phytoextraction

Phytoextraction, also referred to as phytoaccumulation, involves the absorption of contaminants by plant roots from soil or water, followed by their translocation to the above-ground parts such as shoots and leaves. This method has proven highly effective in remediating a wide variety of environmental contaminants, including heavy metals, radionuclides, and certain organic pollutants. As a result, phytoextraction is a versatile and valuable tool for environmental restoration [7].

Phytoextraction is an effective method for removing heavy metals such as lead (Pb), cadmium (Cd), zinc (Zn), and nickel (Ni) from contaminated soils. Hyperaccumulator plants absorb and concentrate these metals in their tissues at levels far higher than those found in the surrounding environment. For example, research has shown that *Cyperus rotundus* efficiently extracts cadmium and lead from soils contaminated by battery scrap waste, making it a promising candidate for bioremediation [8]. Similarly, *Sedum plumbizincicola* has successfully extracted cadmium without increasing its leaching into agricultural soils [9]. Additionally, species such *as Erythrina polymnioides* and *Miconia* from the Ecuadorian Amazon have demonstrated the ability to extract cadmium and zinc, further illustrating the versatility of phytoextraction techniques [10].

The effectiveness of phytoextraction is influenced by several factors, including the bioavailability of contaminants, soil conditions, and the plant species involved. To enhance metal uptake by plant roots, soil amendments such as chelating agents or organic matter are often applied to increase the bioavailability of metals. Once the contaminants are absorbed and stored in the plant tissues, the biomass is harvested and either safely disposed of or processed to recover valuable metals through a method known as phytomining [11]. This approach not only helps remediate contaminated soils but also offers the potential to provide secondary raw materials for industrial applications [12].

2.2 Phytostabilization

Phytostabilization is a process by which plants reduce the mobility or bioavailability of contaminants in soil, thereby preventing their spread through water or air. This mechanism is particularly effective in managing heavy metal contamination in soils where the extraction of pollutants is either impractical or undesirable. In phytostabilization, plants immobilize contaminants by adsorbing them onto their root surfaces, precipitating them within the root zone, or altering soil chemistry to decrease contaminant solubility [13, 14].

Various plant species with extensive root systems are used in phytostabilization to effectively immobilize heavy metals and prevent their migration into surrounding environments. Commonly used species include *Vetiveria zizanioides* (vetiver grass) and *Cyperus alternifolius*, known for their tolerance to contaminated soils and ability to stabilize metals such as lead, cadmium, and zinc within the root zone [15]. These plants work by adsorbing contaminants onto their root surfaces, altering the soil chemistry, and trapping pollutants within the rhizosphere, preventing further contamination of groundwater and air. The selection of plant species for phytostabilization depends on specific site conditions, contaminant types, and environmental factors, all of which can influence the effectiveness of the remediation process. This method is particularly beneficial in areas at risk of erosion or leaching, as it creates vegetative cover that not only prevents soil erosion but also minimizes the likelihood of contaminants being transported off-site.

2.3 Phytodegradation

Phytodegradation, also referred to as phytotransformation, is the process by which plants absorb and break down contaminants through their metabolic pathways. Complex pollutants are converted into simpler, less harmful compounds that are then integrated into plant tissues to support growth. This transformation is facilitated by enzymes such as dehalogenases and oxygenases, which are either released within the plant tissues or in the rhizosphere. Phytodegradation is particularly effective for the treatment of chemical pollutants, including chlorinated solvents, herbicides, and hydrocarbons in soil, groundwater, and air. The success of this process depends on the plant's capacity to absorb contaminants and convert them into non-toxic or less harmful metabolites [16]. When phytodegradation occurs in the rhizosphere, plants release specific enzymes from their roots to degrade pollutants before absorption, further enhancing the efficiency of contaminant breakdown [5].

A recent study investigated the ability of *Cyperus papyrus*, a wetland plant, to metabolize neonicotinoid insecticides. The plant absorbed these chemicals through its roots, transporting them to the shoots, where they were broken down through processes such as nitro reduction, hydroxylation, and demethylation. The study identified key enzymes and genes involved in the plant's response to neonicotinoid exposure, highlighting its potential for removing these contaminants in constructed wetlands [17].

2.4 Rhizofiltration

Rhizofiltration involves the use of plant roots to absorb, adsorb, or precipitate contaminants from water [18, 19]. This method is particularly effective for treating contaminated water sources such as industrial wastewater, stormwater runoff, and polluted groundwater. Rhizofiltration is especially efficient at removing heavy metals and radionuclides from water, and can also address certain organic contaminants [20]. The process works primarily by utilizing plant root systems as a natural filter, where pollutants either bind to the root surfaces or are taken up by the plant for further processing.

Plants like *Helianthus annuus* and *Eichhornia crassipes* are commonly used in rhizofiltration due to their extensive root systems and a high tolerance for pollutants [4]. This technique can be implemented in constructed wetlands or applied in situ to treat contaminated water bodies, offering a cost-effective and natural alternative to traditional filtration methods.

2.5 Phytovolatilization

Phytovolatilization is a process within phytoremediation in which plants absorb contaminants through their roots and release them into the atmosphere in a volatile form through transpiration. This method is particularly effective for certain heavy metals and metalloids, such as mercury, selenium, and arsenic. During this process, plants uptake contaminants from soil or water, translocate them through their tissues, and convert them into less harmful or volatile compounds. These compounds are then released as gases into the atmosphere. For example, *Pteris vittata* has demonstrated the effective removal of arsenic from contaminated soils through phytovolatilization, achieving approximately 90% reduction [21].

One of the main advantages of phytovolatilization is its ability to address widespread contamination, including in remote or difficult-to-access areas. Certain plant species, such as Chinese brake fern (*Pteris vittata*) and *Cyrtomium macrophyllum*, have demonstrated significant potential for mercury phytovolatilization due to their high mercury accumulation capabilities and resistance to mercury stress. These species exhibit high translocation factors, making them suitable candidates for phytoremediation efforts in contaminated environments [22]. However, phytovolatilization poses challenges, particularly the risk of secondary contamination due to the release of volatile pollutants into the atmosphere, which may impact surrounding areas. This concern highlights the importance of carefully selecting plant species and conducting proper monitoring to ensure that volatilized pollutants do not cause further environmental harm [23].

2.6 Translocation and accumulation processes

Translocation and accumulation are vital processes in phytoremediation, where plants absorb contaminants through their roots and transport them to above-ground tissues. These processes are influenced by several factors, including the plant species, the type of contaminant, and the surrounding environmental conditions.

For instance, *Phragmites australis* has demonstrated the ability to effectively translocate and accumulate heavy metals such as copper (Cu) and zinc (Zn), making it suitable for remediating contaminated soils [24]. The effectiveness of a plant in accumulating contaminants is often measured using translocation factors (TF) and bioconcentration factors (BCF), with higher values indicating better performance in pollutant extraction and translocation.

Furthermore, plants like *Rhizophora apiculata* exhibit varying translocation behaviors depending on the metal involved. For example, it efficiently translocates lead (Pb) ($TF > 1$), but shows lower translocation efficiency for copper ($TF < 1$), suggesting that it can play dual roles in both phytoextraction and phytostabilization [25]. Understanding the combination of translocation and accumulation factors is crucial for assessing a plant's potential in phytoremediation, ensuring the successful remediation of contaminated sites.

3. Selection of plant species in phytoremediation system

The success of a phytoremediation system largely depends on the careful selection of plant species, as different plants vary in their ability to tolerate, accumulate, or degrade specific pollutants. The selection process must consider the plant's physiological characteristics, ecological adaptability, and pollutant-specific mechanisms to ensure optimal remediation outcomes.

3.1 Criteria for plant selection

The appropriate selection of plant species is crucial for the success of phytoremediation projects. Several factors must be considered to ensure the plant's ability to survive in contaminated environments, extract pollutants, and remediate the site. The following criteria are generally used for plant selection:

3.1.1 Contaminant tolerance

Plants must exhibit high tolerance to pollutants and the ability to thrive in contaminated environments. Species such as *Typha angustifolia* and *Juncus effusus* are frequently chosen due to their demonstrated tolerance to micropollutants and superior ability to absorb and accumulate contaminants [26]. Additionally, plants with large biomass production, such as *Vetiveria zizanioides*, are favored for their capacity to take up heavy metals while maintaining growth in contaminated soils [27].

Selecting species based on their tolerance to specific contaminants, such as heavy metals or organic pollutants, also ensures more targeted remediation efforts. For instance, sweet potato cultivars have been shown to tolerate high levels of lead (Pb) contamination, demonstrating efficient uptake without significant growth inhibition [28]. Plant selection, therefore, together with pollutant tolerance, biomass production, and the capacity for pollutant accumulation, ensures both ecological sustainability and effective remediation outcomes.

3.1.2 Root structure and depth

Root structure is a critical factor in selecting plant species for phytoremediation, as it directly affects the plant's capacity to absorb, translocate, and accumulate contaminants. Plants with extensive, fibrous root systems, such as *Vetiveria zizanioides* and *Phragmites australis*, are often favored for their ability to penetrate the soil more effectively and enhance pollutant uptake [29]. Additionally, the root system must demonstrate a high tolerance to contaminants and be capable of forming symbiotic relationships with soil microbes, which further support pollutant degradation and stabilization [30].

In urban and industrial environments, Populus balsamifera (balsam poplar) proves to be particularly effective in phytoremediation due to its robust root system, which enhances both pollutant absorption and soil stabilization. Studies have shown that while contamination levels can shape the microbial community associated with the roots of other plants, the fungal community associated with Populus balsamifera remains resilient, making it an ideal candidate for such efforts in polluted environments [31].

3.1.3 Biomass production

Biomass production is another important feature for plant selection, particularly for species used in large-scale applications. High biomass-producing plants, such as *Helianthus annuus* (sunflower) and *Arundo donax* (giant reed), are favored because they can absorb significant amounts of contaminants while generating substantial biomass, which in turn can be utilized for bioenergy production [29].

Additionally, plants with high biomass output, such as *Salix viminalis* (willow) and *Eichhornia crassipes* (water hyacinth), can simultaneously improve soil or water quality and provide renewable energy sources [32]. The dual-purpose use of these plants, both for contamination clean-up and biomass generation, enhances the overall sustainability of the remediation process, making it a valuable criterion in plant selection.

3.1.4 Adaptability to local conditions

In phytoremediation, selecting plant species that can adapt to local environmental conditions is crucial for ensuring the success of the remediation process. Plants that are well-suited to the local climate, soil composition, and pollutant levels are more likely to thrive and effectively accumulate or degrade contaminants. For instance, *Pinus halepensis* has shown remarkable adaptability to contaminated soils, maintaining growth and survival rates even in environments with multiple heavy metal pollutants [33]. Similarly, *Dittrichia viscosa* demonstrates tolerance to both drought and metal pollutants such as arsenic and cadmium, making it a promising candidate for phytoremediation in harsh environments [34].

The ability of plants to adapt to the specific conditions of a contaminated site reduces the need for external inputs, such as water and fertilizers, thereby enhancing the sustainability of the remediation process. Locally adapted species also tend to establish stronger associations with indigenous microbial communities, further improving the efficiency of phytoremediation [35].

3.1.5 Contaminant-specific uptake mechanisms

In phytoremediation, contaminant-specific uptake mechanisms are crucial in determining a plant's ability to absorb and detoxify particular pollutants. For heavy metals such as lead, cadmium, and zinc, plants utilize specific transport proteins and enzymes to bind and translocate these metals into their tissues. This process often involves chelation, where metals are bound to organic molecules, followed by compartmentalization within vacuoles to prevent toxicity [36]. For organic pollutants, plants employ enzymatic pathways, such as those involving cytochrome P450 enzymes, to break down contaminants into less harmful compounds, which can either be incorporated into the plant biomass or volatilized into the atmosphere [37].

These mechanisms vary based on the type of contaminant and the plant species involved. Hyperaccumulator plants, for instance, have evolved to efficiently translocate and store metals in their shoots, making them ideal for phytoextraction. Conversely, plants used for phytodegradation excel at transforming organic pollutants into non-toxic substances through enzymatic activity in their roots or leaves. This adaptability enables the implementation of targeted phytoremediation strategies tailored to specific contaminants and environmental conditions.

3.1.6 Non-invasiveness and ecological impact

When selecting plants for phytoremediation, it is essential to ensure that they do not pose an invasive risk to the local ecosystem. Invasive species can outcompete native plants, disrupt biodiversity, and create long-term ecological imbalances. Therefore, when non-native species are used, monitoring and control measures must be in place to prevent their spread beyond the remediation site. Phytoremediation, as a whole, is a non-invasive and environmentally friendly approach, utilizing plants to remove contaminants without causing significant disturbance to the ecosystem. By enhancing soil quality, conserving biodiversity, and contributing to habitat restoration, it offers a sustainable alternative to conventional remediation methods. Its careful integration into diverse ecosystems can minimize ecological impact and promote long-term environmental health [38].

3.2 Hyperaccumulator plants

Hyperaccumulator plants are species with a unique ability to absorb and accumulate unusually high concentrations of metals or other contaminants in their tissues. These plants are particularly valuable in phytoextraction, where the goal is to remove metals from contaminated soils or water. Hyperaccumulators can store metals at levels that would typically be toxic to most plants, often concentrating pollutants in their shoots and leaves. They possess specialized biochemical mechanisms that allow them to sequester metals in vacuoles or cell walls, thereby avoiding toxic effects. For instance, hyperaccumulators produce metal-binding proteins like metallothionein and phytochelatin, which help detoxify absorbed metals. These plants also thrive in nutrient-poor, metal-rich soils where other species would struggle to survive [39].

Numerous hyperaccumulator species have been identified and are widely used in phytoremediation, as shown in **Table 1**. *Pteridium aquilinum* is a highly effective hyperaccumulator of hexavalent chromium (Cr), capable of accumulating up to 11,854 $mg.kg^{-1}$ of Cr in its underground parts. This fern is particularly well-suited for remediating chromium-contaminated environments, especially in industrial and mining areas where hexavalent chromium poses significant ecological risks. Its ability to sequester such high concentrations of Cr makes it an invaluable tool for phytoremediation projects targeting heavy metal contamination [40].

Similarly, *Acorus calamus* is known for its ability to accumulate copper (Cu) from contaminated water sources. Frequently used in aquatic phytoremediation, this plant

Plant species	Accumulated contaminant	Accumulation capabilities	Application	Reference
Pteridium aquilinum	Hexavalent chromium (Cr)	11,854 $mg.kg^{-1}$ in underground parts	Industrial and mining areas with chromium contamination	[40]
Acorus calamus	Copper (Cu)	High tolerance and bioaccumulation in aquatic settings	Wetlands and constructed water treatment systems for copper removal	[41]
Pistia stratiotes	Copper (Cu)	Large root system, high bioaccumulation factor	Constructed wetlands and water treatment systems for copper removal	[41]
Azolla pinnata	Copper (Cu)	High bioaccumulation factor, effective in low copper concentrations	Aquatic settings for copper detoxification	[41]
Arabis paniculata	Lead (Pb), Zinc (Zn), Cadmium (Cd)	High metal tolerance and accumulation, suited for phytoextraction	Industrial and mining sites for heavy metal-contaminated soil	[42]

Table 1.
Summary of the species, their accumulation capabilities, and their respective applications.

demonstrates strong bioaccumulation capabilities and a high tolerance to metal pollution. It is commonly used in wetlands and constructed water treatment systems as part of an eco-friendly approach to mitigating metal pollutants [41].

Pistia stratiotes, commonly known as water lettuce, is another efficient hyperaccumulator of copper (Cu). Noted for its large root system and high bioaccumulation factor, this plant is widely employed in constructed wetlands and water treatment systems to address copper contamination, providing a sustainable solution for managing metal pollutants in water [41]. *Azolla pinnata* is also highly effective in absorbing copper in aquatic environments, thanks to its high bioaccumulation factor and rapid growth rate, which allows it to quickly colonize contaminated water bodies [41]. Additionally, *Arabis paniculata* is recognized for its tolerance to and accumulation of heavy metals such as lead (Pb), zinc (Zn), and cadmium (Cd). This species is frequently used in phytoremediation projects to restore soils contaminated by industrial and mining activities due to its high metal tolerance and accumulation capacity [42].

Although hyperaccumulators show great potential for remediating metal-contaminated soils, they have certain limitations. These plants typically produce low biomass, which slows the rate at which they accumulate contaminants compared to faster-growing species. Additionally, hyperaccumulators tend to be specialized for specific metals, limiting their use to certain contaminants. As a result, in large-scale remediation projects, hyperaccumulators are often combined with other species that can target a wider range of contaminants or produce more biomass [43].

3.3 Genetic and biotechnological enhancements in phytoremediation

Recent advancements in genetic engineering and biotechnology have greatly enhanced the potential of phytoremediation. These innovations allow plants to remediate a broader range of contaminants more effectively and improve their ability to withstand harsh environmental conditions, such as high pollution levels or extreme soil environments. This section will explore key biotechnological approaches, including genetic modifications, CRISPR technology, and the use of plant-associated microorganisms.

3.3.1 Genetically modified plants for phytoremediation

Genetically modified plants (GMPs) have shown great promise in enhancing the effectiveness of phytoremediation. By introducing specific genes that improve the uptake and detoxification of heavy metals and other pollutants, these transgenic plants can target environmental contaminants more efficiently. For example, advancements in CRISPR-Cas9 technology have facilitated the genetic modification of ornamental and aromatic plants, increasing their tolerance and capacity to remove heavy metals from polluted environments [44]. This biotechnological approach represents a significant breakthrough, allowing scientists to develop plant species with enhanced detoxification abilities [45].

In another study, CRISPR-Cas9 technology was used to modify tobacco plants by eliminating xyloglucan from their cell walls. This modification improved the plant's ability to retain arsenic in its roots while minimizing translocation to other parts of the plant, making it more effective in remediating arsenic-contaminated soils [42]. These advancements demonstrate the potential of CRISPR-Cas9 to create plants specifically designed for large-scale environmental restoration efforts.

3.3.2 Plant-microbe symbiosis and rhizosphere engineering

Plant-microbe symbiosis and rhizosphere engineering are essential for improving the efficiency of phytoremediation. Plant growth-promoting bacteria (PGPR), such as Pseudomonas and Bacillus, significantly enhance the ability of hyperaccumulator plants like *Helianthus annuus* to absorb and accumulate heavy metals such as zinc (Zn) and cadmium (Cd). These beneficial microbes interact with the plant's root system, modifying the rhizosphere by increasing nutrient availability and enhancing metal solubility, thus facilitating greater metal uptake [46]. The rhizosphere, the biologically active zone around plant roots, becomes a focal point for pollutant degradation when inoculated with these microbes.

Rhizosphere engineering involves deliberately manipulating this zone by introducing specific microbes or altering soil conditions to boost microbial activity. This approach enhances the plant's natural remediation capabilities by promoting processes like metal chelation, organic pollutant degradation, and nutrient cycling. Studies have shown that microbial inoculants can increase phytoremediation efficiency, reduce toxicity, and improve plant growth in contaminated environments [47].

3.3.3 Integration of phytoremediation with synthetic biology

The integration of synthetic biology into phytoremediation offers exciting new possibilities for enhancing the removal of contaminants from the environment. Synthetic biology enables the engineering of both plants and microbial communities to improve their capabilities for pollutant uptake, degradation, and detoxification. For example, modifying rhizosphere microorganisms using synthetic biology techniques can enhance interactions between plants and microbes, optimizing the rhizosphere environment for more efficient phytoremediation [48]. By designing microbial strains or plant genes specifically tailored to target certain pollutants, scientists can develop systems that precisely degrade contaminants.

This integration also enables the construction of synthetic regulatory networks in plants, allowing them to respond to environmental signals and dynamically control the expression of remediation genes. These innovations present a promising approach for addressing complex environmental pollution in a targeted and sustainable way, significantly improving the effectiveness of phytoremediation in large-scale applications.

4. Applications of phytoremediation in environmental remediation

Phytoremediation has a wide range of applications. Depending on the type of pollutant and the medium impacted, different strategies can be employed in various environmental settings involving soil and water treatment as well as air pollution control.

4.1 Soil remediation

Phytoremediation in soil remediation presents an eco-friendly and cost-effective solution to remove heavy metals and other pollutants. This method utilizes plants, often combined with soil microbes, to absorb, translocate, and stabilize contaminants in the soil. Techniques like phytoextraction and phytostabilization are commonly used, depending on the type of pollutant and soil conditions. For instance, hyperaccumulator species like *Phragmites australis* have shown high efficiency in removing metals such as zinc and cadmium from contaminated soils, especially in industrial areas [49]. Another example is *Helianthus annuus* (sunflower), widely used for phytoextraction due to its ability to accumulate lead and other toxic metals from the soil [50].

Many large-scale phytoremediation projects have integrated innovative techniques, such as bioaugmentation and biochar amendments, to enhance soil fertility and contaminant removal. Research has shown that combining biochar with plants like *Jatropha curcas* significantly reduces the bioavailability of heavy metals while improving soil quality [51]. These advancements emphasize phytoremediation's potential to tackle severe soil pollution in mining areas and urban settings, providing a sustainable alternative to conventional remediation methods.

4.2 Water and wetland treatment

Phytoremediation has proven to be an effective approach for treating water and wetlands, particularly for removing pollutants like heavy metals and organic compounds from contaminated water bodies. Constructed wetlands utilizing plants such as *Phragmites australis* and *Typha latifolia* have been successfully implemented to treat wastewater reducing contaminants like nitrates, phosphates, and heavy metals [52]. These systems leverage the natural capacity of plants to absorb, accumulate, and detoxify pollutants, providing a sustainable and cost-effective alternative to traditional water treatment methods.

Recent studies have also emphasized the potential of floating wetland systems using terrestrial plant species such as *Canna indica* and *Chrysopogon zizanioides*, which have demonstrated high pollutant removal efficiencies in treating domestic sewage [53]. Additionally, constructed wetlands utilizing Napier grass have shown remarkable chromium removal from industrial wastewater, with removal capacities as high as 99.96% [54].

4.3 Air pollution mitigation

Phytoremediation has also proven to be an effective method for mitigating air pollution by utilizing the natural abilities of plants to absorb, adsorb, and degrade airborne contaminants. Plants such as *Ficus religiosa* and *Mangifera indica* have been identified for their high tolerance to air pollution and are commonly used to reduce harmful pollutants, including particulate matter (PM), nitrogen oxides (NOx), and sulfur dioxide (SO_2), in urban environments [55]. These plants act as biofilters, removing pollutants from the air while providing ecological benefits such as improved air quality and reduced urban heat island effects.

Advances in the use of green walls and vertical gardens have further enhanced the application of phytoremediation in urban areas. These systems not only improve air purification but also process larger volumes of air, making them suitable for areas with high pollution levels [56]. Studies show that broad-leaf species like white mulberry and plane trees are particularly effective in reducing pollutants such as ozone (O3) and carbon monoxide (CO), making phytoremediation a valuable strategy for air pollution control in cities [57].

4.4 Urban and industrial sites

Phytoremediation has been successfully applied in urban and industrial areas to address contamination from heavy metals, hydrocarbons, and other pollutants. This method provides a sustainable and cost-effective solution for cleaning up sites impacted by years of industrial activity, especially where metals like lead, cadmium, and copper have accumulated. For example, *Ulmus glabra* has demonstrated a high capacity for heavy metal accumulation, making it particularly valuable in contaminated urban and industrial zones [58]. Tree-based phytoremediation is often preferred in urban environments as it not only mitigates soil contamination but also offers additional benefits such as air purification and erosion control [59].

In areas where both soil remediation and biomass production are priorities, combining phytoremediation with biomass cultivation has proven to be an effective strategy. For instance, studies have shown that in industrial sites, this approach not only reduced hydrocarbon concentrations but also generated viable biomass suitable for bioenergy applications [60]. This dual-purpose technique is gaining recognition as a promising way to restore contaminated urban and industrial areas while producing renewable resources.

5. Advantages and limitations of phytoremediation

Phytoremediation is a promising and sustainable method for cleaning contaminated environments, offering significant ecological, economic, and social benefits. However, like any remediation technique, it has inherent limitations that must be considered for effective use. A thorough understanding of both the advantages and challenges of phytoremediation is essential for its successful application.

5.1 Environmental and ecological benefits

One of the most significant advantages of phytoremediation is its environmentally friendly approach to remediation. Unlike traditional methods that rely on mechanical, chemical, or thermal interventions, phytoremediation works in harmony with natural processes to restore contaminated environments. The use of plants not only aids in pollutant removal but also contributes to the overall improvement of soil, water, and air quality.

5.1.1 Soil health improvement

Phytoremediation significantly enhances soil health and provides various environmental and ecological benefits. This method utilizes plants to detoxify and immobilize contaminants, thereby improving the physical, biological, and chemical properties of soil. The reduction of heavy metals in the soil enhances soil fertility and promotes microbial activity [6]. Phytoremediation facilitates pollutant removal and promotes long-term soil recovery, mitigating erosion and runoff, thereby enhancing overall ecological stability [61]. Hyperaccumulator plants introduced in contaminated soils enhance soil structure restoration and facilitate nutrient cycling, resulting in sustainable improvements in soil health [16].

5.1.2 Biodiversity conservation

Phytoremediation is essential for biodiversity conservation as it rehabilitates contaminated environments and promotes the growth of native plant species. This method effectively eliminates harmful pollutants while simultaneously fostering conditions conducive to the flourishing of diverse ecosystems. Halophytes exhibit high tolerance to salt and pollutants, playing a crucial role in phytoremediation and biodiversity restoration. Their significance lies in addressing ecosystem degradation and enhancing food security and renewable resources [62]. These plants offer the dual advantage of environmental detoxification and the enhancement of native species diversity.

Agroforestry systems integrate trees and other plants with agriculture, thereby enhancing biodiversity conservation via phytoremediation. This method enhances soil health, augments microbial diversity, and facilitates intricate species interactions, thereby increasing biodiversity and ecosystem productivity [63]. The multi-species planting model of agroforestry improves soil quality and supports a greater diversity of plant and animal species relative to monoculture systems, positioning it as a sustainable approach for remediation and biodiversity conservation.

5.1.3 Carbon sequestration and climate change mitigation

Phytoremediation plays a crucial role in carbon sequestration and addressing climate change by harnessing the inherent capacity of plants to absorb atmospheric CO_2 and transform it into biomass. *Phragmites australis* is utilized in urban forestry and constructed wetlands, contributing to carbon sequestration, enhancing air quality, and providing habitats for wildlife [64]. The establishment of greenbelts and the rehabilitation of polluted areas highlight the significant contributions of these plants to environmental restoration and climate management, positioning phytoremediation as a crucial strategy for meeting climate change objectives.

Furthermore, agroforestry systems, especially those that combine tree species with crops, demonstrate significant potential for carbon sequestration, while also enhancing biodiversity and soil health. Agroforestry initiatives in semi-arid areas have shown potential in enhancing biomass and carbon stock, thereby playing a role in climate change mitigation and promoting agricultural sustainability [65].

5.2 Economic benefits

Phytoremediation typically presents a more economical alternative to conventional remediation methods, which require costly mechanical procedures such as excavation, transportation, and disposal of contaminated soil, or the application of intricate chemical treatments. The reduced expense associated with phytoremediation renders it a viable choice, especially for large-scale projects or in low- to middle-income areas where financial constraints are prevalent.

5.2.1 Cost-effectiveness

Phytoremediation is acknowledged for its economic advantages, positioning it as an attractive option for addressing contaminated soils and water bodies. This sustainable method utilizes the inherent capacity of plants to take in, break down, or stabilize pollutants, eliminating the necessity for expensive mechanical or chemical interventions. Research indicates that hyperaccumulator plants like *Vetiveria zizanioides* and *Phragmites australis* can considerably lower remediation expenses while ensuring effective metal uptake and water treatment [66, 67]. Furthermore, the integration of phytoremediation with biochar or microbial inoculation significantly boosts its economic feasibility by enhancing soil fertility and contaminant extraction, thereby providing both environmental and monetary advantages.

Compared to conventional approaches, phytoremediation demands less energy, labor, and materials, positioning it as a more sustainable long-term strategy for the recovery of contaminated sites. For example, the application of constructed wetlands utilizing *Typha angustifolia* for the treatment of acid mine drainage has demonstrated significant effectiveness and cost-efficiency, with removal rates reaching 94.35% for iron and 85.21% for manganese, all while circumventing the elevated expenses linked to sludge generation in active processing techniques [68]. Phytoremediation serves as an essential strategy in extensive environmental management, particularly in resource-limited contexts.

5.2.2 Applicability to large-scale and low-resource settings

Phytoremediation demonstrates significant potential for implementation in extensive and resource-limited environments, largely because it utilizes natural mechanisms and requires minimal infrastructure. This method enables economical restoration of polluted soils and water across large regions, eliminating the requirement for advanced machinery or chemical solutions. For instance, rapidly proliferating species such as *Helianthus annuus* (sunflower) and *Eichhornia crassipes* (water hyacinth) are frequently employed in extensive initiatives due to their capacity to uptake heavy metals and various pollutants, all while necessitating minimal upkeep [69].

In low-resource settings, phytoremediation can be applied with minimal financial investment, presenting a viable solution for countries or regions facing resource limitations. Utilizing local plant species significantly improves the adaptability of this method, guaranteeing successful results across diverse climatic conditions and soil types [65].

5.3 Social and Esthetic benefits

Phytoremediation presents considerable social and esthetic advantages, especially in urban settings. The inclusion of green spaces and vegetation in urban areas significantly contributes to better air quality while simultaneously elevating the esthetic value of cityscapes. Moreover, integrating phytoremediation into public areas can enhance community involvement and promote environmental consciousness.

5.3.1 Urban beautification and green spaces

Phytoremediation offers notable social and esthetic advantages by improving urban landscapes and encouraging the development of green areas. In urban areas, the use of plants for remediation enhances soil and environmental quality, leading to the development of more appealing and habitable spaces. Ornamental trees and plants, commonly utilized in phytoremediation, play a dual role by alleviating heavy metal pollution while simultaneously enhancing green spaces in densely populated regions, thus aiding in urban beautification [70]. The dual functionality of phytoremediation serves as an important tool in urban planning, enhancing environmental sustainability and elevating the esthetic value of urban areas.

Furthermore, the implementation of green infrastructure, including vertical gardens and green roofs in urban settings, highlights the architectural and ecological advantages associated with phytoremediation. These systems improve the esthetics of buildings, reduce the urban heat island effect, and aid in enhancing air quality, thereby rendering cities healthier and more visually attractive [71]. Phytoremediation serves as a remediation strategy while simultaneously providing cultural ecosystem services that improve social well-being and elevate the quality of life in urban environments.

5.3.2 Community engagement and education

Phytoremediation encourages community involvement and learning by engaging local residents in environmental restoration initiatives, increasing awareness of pollution challenges, and advocating for sustainable practices. Communities can engage in the planting and upkeep of phytoremediation sites, promoting a sense of ownership and commitment to environmental care. Programs associated with these initiatives assist communities in taking the scientific principles of phytoremediation and its advantages for public health and ecological sustainability [72].

Furthermore, phytoremediation efforts create possibilities for educational institutions to integrate practical learning experiences into their programs. Through the involvement of students and local stakeholders in remediation activities, these initiatives improve environmental education and encourage individuals with the knowledge needed to address future environmental challenges. This collaborative effort enhances local ecosystems while supporting community connections and encouraging sustained responsibility for the environment.

5.4 Limitations and challenges of phytoremediation

Phytoremediation presents various advantages; however, considerable challenges exist that require to be addressed to further improve its effectiveness and broader implementation. The constraints include the timeframe needed for remediation, the bioavailability of contaminants, reliance on climatic conditions, and the necessity for effective site management.

5.4.1 Long timeframes

Phytoremediation is constrained by the extended timescales necessary for effective remediation, due to the slower growth cycles of plants. This prolonged length creates challenges, particularly for emergencies demanding rapid remediation, because the entire process can take months or years [73]. Additionally, factors such as soil toxicity and adverse environmental conditions may hinder plant development, hence delaying the clean-up process. Improvements in plant breeding and genetic engineering are being investigated to improve growth rates and pollution absorption, hence reducing these time periods [27].

5.4.2 Contaminant bioavailability

A significant restriction of phytoremediation is the difficulty related to the bioavailability of soil pollutants. The efficacy of phytoremediation is greatly affected by the bioavailability of heavy metals in the soil and the capacity of plants to absorb these pollutants. Enhancing metal bioavailability requires optimization of soil conditions and the selection of appropriate plant species. Furthermore, using chelating chemicals such as ethylenediaminetetraacetic acid (EDTA) or citric acid may enhance metal absorption by plant roots; yet, this method may elevate the possibility of metal leaching into groundwater, hence posing environmental hazards [49].

5.4.3 Climatic and environmental limitations

Environmental vulnerability, including severe weather conditions like drought or temperature variations, may adversely affect plant growth, nutrient absorption, and the total bioavailability of soil pollutants [74]. In arid regions such as those in Chile, the severe scarcity of water could hinder the survival of plants and the remediation process, affecting the setting up and ongoing operation of phytoremediation schemes [75]. Moreover, seasonal fluctuations and the biological needs of plants, including particular growth circumstances, may extend the remediation period, making the method less efficient compared to other approaches [73].

5.4.4 Potential for contaminant Re-entry into the ecosystem

The potential of pollutant re-entry into the environment is an important challenge to the sustained efficacy of phytoremediation activities. Upon the breakdown of phytoremediation plants, pollutants like heavy metals or organic pollutants may be reintroduced into the soil or water, thereby affecting the remediation processes. This issue has been highlighted in research on polluted mining locations, where inadequate disposal or degradation of plants resulted in the re-emission of contaminants [76].

Moreover, adverse environmental conditions, such as intense precipitation or soil erosion, may result in the remobilization of contaminants, particularly in areas where phytostabilization is used to isolate pollutants rather than eliminate them. Effective post-remediation techniques, such as the safe removal of biomass and continuous site monitoring, are crucial to reduce the risk of pollutant re-entry and ensure long-term success [77].

5.5 Addressing the challenges of phytoremediation

Research and innovative approaches are being explored to address the problems related to phytoremediation. Multiple approaches have been established that reduce limitations and improve the application of phytoremediation.

5.5.1 Use of soil amendments

Soil amendments, such as biochar, organic matter, or synthetic chelating agents, may be used to enhance the bioavailability of pollutants. Biochar has shown the capacity to increase the bioavailability of heavy metals and improve soil structure, allowing more effective absorption by plant roots. Still, it is necessary to evaluate the advantages of these adjustments against potential environmental consequences [43].

5.5.2 Biotechnology and genetic engineering

Advancements in genetic engineering, particularly the use of CRISPR-Cas9 technology, have made it possible to enhance the tolerance of plants to contaminants, increase their uptake capacity, and even modify plants to break down organic pollutants. By genetically modifying plants to express specific enzymes or metal transporters, innovative approaches are being explored to produce more resilient species that can thrive in harsh environmental conditions and accelerate the phytoremediation process [78].

5.5.3 Combining phytoremediation with other techniques

Combining phytoremediation with other remediation methods may greatly increase its efficacy in addressing the problems with it. For example, integrating phytoremediation with stabilization methods, including the application of amendments such as steel slag or pyrolusite, can mitigate the spread of contaminants like cadmium and arsenic, providing a more holistic approach [79]. This combination enhances the immobilization of contaminants and enhances both the economic and technical viability of the remediation process.

Furthermore, the integration of bio-combined techniques, including the use of plant growth-promoting rhizobacteria or microbial inoculants, significantly enhances the efficiency of phytoremediation. These microorganisms contribute to better metal absorption and reduce stress in plants, thereby increasing their resilience in polluted environments. This combined method has demonstrated greater efficacy in eliminating trace elements compared to phytoremediation used alone [80].

By addressing these challenges with innovative techniques and integrated approaches, phytoremediation has the potential to emerge as a more viable and effective solution for environmental restoration. As studies advance and innovative technologies develop, the promise of phytoremediation to aid in sustainable environmental management expands, presenting reassurance for the remediation of polluted areas across the world.

6. Conclusion

Phytoremediation represents a novel and economically viable method for addressing pollution, employing the inherent capabilities of plants to restore contaminated ecosystems. Plants utilize mechanisms including phytoextraction, phytostabilization, phytodegradation, and rhizofiltration to effectively mitigate various pollutants, such as heavy metals, organic compounds, and radionuclides. Phytoremediation is applicable across various environmental contexts, including industrial sites, agricultural lands, and urban areas. Case studies from various regions have shown its effectiveness, offering important insights for upcoming projects. Furthermore, advancements in biotechnology, such as genetic modifications, CRISPR-Cas9 technology, and plant-microbe symbiosis, have particularly improved the efficiency of phytoremediation. Recent advancements have expanded the potential of phytoremediation by boosting plant tolerance to pollutants, improving contaminant uptake, and facilitating the degradation of complex pollutants that used to be difficult to manage.

Phytoremediation, while advantageous, encounters challenges including extended remediation periods, dependency on climate variables, and limitations in addressing non-bioavailable contaminants. Ongoing research and innovations in synthetic biology, along with improved integration with other remediation techniques, are expected to address these limitations. Future research should prioritize the optimization of genetic enhancements, the advancement of rhizosphere engineering, and the integration of real-time monitoring technologies, such as machine learning, to enhance the prediction of remediation outcomes and expedite the process.

Phytoremediation provides interesting ecological and economic benefits alongside its technical advantages. This process enhances biodiversity, promotes soil health, and facilitates carbon sequestration, thereby aiding in comprehensive environmental restoration initiatives. Phytoremediation is more cost-effective than conventional remediation methods, rendering it advantageous for extensive projects and areas with constrained financial resources. The integration with global environmental policies, including the Sustainable Development Goals (SDGs) and the Paris Agreement, underscores its significance in addressing pollution and climate change on a global scale.

Phytoremediation presents significant potential for sustainable environmental management in the future. This approach addresses current pollution issues while also serving as a proactive solution that aligns with global goals for the environment. Greater funding for research, policy support, and public awareness is crucial for comprehending the full potential of phytoremediation. Collaboration among scientists, policymakers, and communities can enhance phytoremediation's role in promoting a healthier, cleaner, and more resilient planet for future generations.

DOI: http://dx.doi.org/10.5772/intechopen.1007671

Author details

Harlina Ahmad
School of Industrial Technology, Universiti Sains Malaysia, Penang, Malaysia

*Address all correspondence to: harlinaa@usm.my

References

[1] Belete MD. Ecohydrological nature-based solution for climate resilience and biodiversity enhancement in water-limited ecosystem: Perspectives and proof of concepts. Ecohydrology and Hydrobiology. 2023;**23**(4):507-517

[2] Ren JWF, Coffman GC. Integrating the resilience concept into ecosystem restoration. Restoration Ecology. 2023;**31**(5):e13907

[3] Khan SN, Mohsin M, Anwar MF. Efficient removal of heavy metals by phytoremediation techniques. Journal Clean WAS (JCleanWAS). 2022;**6**(2):62-65

[4] Sophia S, Shetty Kodialbail V. Phytoremediation of Soil for Metal and Organic Pollutant Removal. Bioprocess Engineering for Bioremediation: Valorization and Management Techniques. 2020, pp. 45-66

[5] Ranjan R, Sow S. The role of plant enzymes in rhizosphere-based phytoremediation. Environmental Biotechnology Review. 2021;**14**(2):55-67

[6] Singh H, Pant G. Phytoremediation: Low input-based ecological approach for sustainable environment. Applied Water Science. 2023;**13**(3):85

[7] Ganesan S, Panda S. Phytoremediation: Sustainable approach for the removal of heavy metals from the environment using plants. In: IOP Conference Series: Materials Science and Engineering. Bristol, United Kingdom: IOP Publishing; 2020. p. 012096

[8] Al-Huqail AA, Taher MA, Širić I, Goala M, Adelodun B, Choi KS, et al. Bioremediation of battery scrap waste contaminated soils using coco grass (Cyperus rotundus L.): A prediction Modeling study for cadmium and Lead Phytoextraction. Agriculture. 2023;**13**(7):1411

[9] Zhou T, Huang H, Mu T, Wang Y, Zhou J, Li X, et al. Does phytoextraction with sedum plumbizincicola increase cadmium leaching from polluted agricultural soil? International Journal of Phytoremediation. 2024;**26**(2): 241-249

[10] Chamba-Eras I, Griffith DM, Kalinhoff C, Ramírez J, Gázquez MJ. Native hyperaccumulator plants with differential phytoremediation potential in an artisanal gold mine of the Ecuadorian Amazon. Plants. 2022;**11**(9):1186

[11] Balint R, Boajă IP. Assisted Phytoextraction as a Nature-Based Solution for the Sustainable Remediation of Metal (loid)-Contaminated Soils. Integrated Environmental Assessment and Management. 2024

[12] Minuț M, Cozma P, Roșca M, Vasilachi IC, Hlihor RM, Diaconu M, et al. Recovery of valuable heavy metals from polluted soil using Phytomining process-a new challenge for earning secondary raw materials and health risk reduction. In: 2021 International Conference on e-Health and Bioengineering (EHB). Romania: IEEE; 2021. pp. 1-4

[13] Hegedus C, Pașcalău SN, Andronie L, Rotaru AS, Cucu AA, Dezmirean DS. The journey of 1000 leagues towards the decontamination of the soil from heavy metals and the impact on the soil–plant–animal–human chain begins with the first step: Phytostabilization/phytoextraction. Agriculture. 2023;**13**(3):735

[14] Rasool FU, Ahmad L, Hassan A, Iqbal S, Sofi MA. Phytoremedation: An effective way to treat heavy metal contamination. A review. Current Journal of Applied Science and Technology. 2023;**42**(47):92-99

[15] Nokande SE, Razavi SM, Mohammadian MA. The capacity of heavy metal remediation by Cyperus alternifolius, Chrysopogon zizanioides (L.) Roberty, and Aloe vera (L.) Burm. f. under industrial and urban wastewater treatment. Chiang Mai University Journal of Natural Sciences. 2022;**21**(4):e2022057

[16] Khan S, Masoodi TH, Pala NA, Murtaza S, Mugloo JA, Sofi PA, et al. Phytoremediation prospects for restoration of contamination in the natural ecosystems. Water. 2023;**15**(8):1498

[17] Liu H, Tang X, Tam NFY, Li Q, Ruan W, Xu X, et al. Phytodegradation of neonicotinoids in Cyperus papyrus from enzymatic and transcriptomic perspectives. Journal of Hazardous Materials. 2024;**462**:132715

[18] Kristanti RA, Ngu WJ, Yuniarto A, Hadibarata T. Rhizofiltration for removal of inorganic and organic pollutants in groundwater: A review. Biointerface Research in Applied Chemistry. 2021;**11**:12326-12347

[19] Singh S, Kumar V, Dhanjal DS, Datta S, Singh S, Singh J. Rhizoremediation of organic and inorganic pollutants: Advances and challenges. In: Bioremediation for Environmental Sustainability. Amsterdam: Elsevier; 2021;**1**:397-420

[20] Yadav AK, Pathak B, Fulekar MH. Rhizofiltration of heavy metals (cadmium, lead and zinc) from fly ash leachates using water hyacinth (Eichhornia crassipes). International Journal of Environment. 2015;**4**(1):179-196

[21] Sakakibara M, Watanabe A, Inoue M, Sano S, Kaise T. Phytoextraction and phytovolatilization of arsenic from As-contaminated soils by Pteris vittata. In: Proceedings of the Annual International Conference on Soils, Sediments, Water and Energy. 2010;**12**(1):26

[22] Su Y, Han FX, Chen J, Sridhar BM, Monts DL. Phytoextraction and accumulation of mercury in three plant species: Indian mustard (Brassica juncea), beard grass (Polypogon monospeliensis), and Chinese brake fern (Pteris vittata). International Journal of Phytoremediation. 2008;**10**(6):547-560

[23] Guarino F, Miranda A, Castiglione S, Cicatelli A. Arsenic phytovolatilization and epigenetic modifications in Arundo donax L. assisted by a PGPR consortium. Chemosphere. 2020;**251**:126310

[24] Chitimus D, Nedeff V, Mosnegutu E, Barsan N, Irimia O, Nedeff F. Studies on the accumulation, translocation, and enrichment capacity of soils and the plant species phragmites australis (common reed) with heavy metals. Sustainability. 2023;**15**(11):8729

[25] Juswardi, Septiantika ST, Triwardana S, Junaidi E, Sarno. Potential of Rhizophora apiculata Blume in phytoremediation of heavy metals Pb and Cu in the mangrove Forest nature reserve East Coast, Alang-Alang Village, Tanjung Jabung Timur, Jambi. International Journal of Environmental Protection. 2023;**5**(2):14130

[26] Lei Y, Carlucci L, Rijnaarts H, Langenhoff A. Phytoremediation of micropollutants by Phragmites australis, Typha angustifolia, and Juncus effuses. International Journal of Phytoremediation. 2023;**25**(1):82-88

[27] Tyagi SK, Kumar P, Tanveer N, Pandey N, Rajput Y. Recent advances in phytoremediation of hazardous substances using plants: A tool for soil reclamation and sustainability. AgroEnvironmental Sustainability. 2023;**1**(2):180-191

[28] Daurov D, Lim YH, Park SU, et al. Selection and characterization of lead-tolerant sweetpotato cultivars for phytoremediation. Plant Biotechnology Reports. 2024;**18**:327-339

[29] Mei X, Wang Y, Li Z, et al. Root-associated microbiota drive phytoremediation strategies to lead of Sonchus Asper (L.) hill as revealed by intercropping-induced modifications of the rhizosphere microbiome. Environmental Science and Pollution Research. 2022;**29**:23026-23040

[30] Bourdel G, Roy-Bolduc A, St-Arnaud M, Hijri M. Concentration of petroleum-hydrocarbon contamination shapes fungal endophytic community structure in plant roots. Frontiers in Microbiology. 2016;**7**:685

[31] Amabogha ON, Garelick H, Jones H, Purchase D. Combining phytoremediation with bioenergy production: Developing a multi-criteria decision matrix for plant species selection. Environmental Science and Pollution Research. 2023;**30**(14):40698-40711

[32] Nahar K, Sunny SA. Co-benefits of Eichhornia Crassipes (water hyacinth) as sustainable biomass for biofuel production and aquatic ecosystem phytoremediation. Fuels. 2024;**5**(3):317-333

[33] Kharazian P, Cappai G, Boi ME, Porceddu M, Piredda M, De Giudici G, et al. Greenhouse investigation on the phytoremediation potential of pioneer tree Pinus halepensis mill. in abandoned mine site. International Journal of Phytoremediation. 2024;**26**(5):773-783

[34] Anglana C, Capaci P, Barozzi F, Migoni D, Rojas M, Stigliano E, et al. Dittrichia viscosa selection strategy based on stress produces stable clonal lines for phytoremediation applications. Plants. 2023;**12**(13):2499

[35] Ariyakanon N. Application of local species for sustainable phytoremediation: 10.32526/ennrj/21/20230125. Environment and Natural Resources Journal. 2023;**21**(5):381-389

[36] Skuza L, Szućko-Kociuba I, Filip E, Bożek I. Natural molecular mechanisms of plant hyperaccumulation and hypertolerance towards heavy metals. International Journal of Molecular Sciences. 2022;**23**(16):9335

[37] Liu X, Yang A, Kümmel S, Richnow HH. Uptake and metabolization of HCH isomers in trees examined over an annual growth period by compound-specific isotope analysis and enantiomer fractionation. Environmental Science and Technology. 2022;**56**(14):10120-10130

[38] Sharma JK, Kumar N, Singh NP, Santal AR. Phytoremediation technologies and their mechanism for removal of heavy metal from contaminated soil: An approach for a sustainable environment. Frontiers in Plant Science. 2023;**14**:1076876

[39] Rai GK, Bhat BA, Mushtaq M, Tariq L, Rai PK, Basu U, et al. Insights into decontamination of soils by phytoremediation: A detailed account on heavy metal toxicity and mitigation strategies. Physiologia Plantarum. 2021;**173**(1):287-304

[40] Eslava-Silva FDJ, Muñíz-Díaz de León ME, Jiménez-Estrada M. Pteridium aquilinum (Dennstaedtiaceae), a novel hyperaccumulator species of hexavalent chromium. Applied Sciences. 2023;**13**(9):5621

[41] Dewi JA, Nurwahyuni I, Munir E. Plant effectiveness of Acorus calamus, Pistia stratiotes, Azolla pinnata as Hyperaccumulator candidate of phytoremediation agent for copper (Cu) absorption. Jurnal Pembelajaran Dan Biologi Nukleus (JPBN). 2023;**9**(2):276-289

[42] Wang M, Song X, Guo S, Li P, Xu Z, Xu H, et al. Using CRISPR-Cas9 technology to eliminate xyloglucan in tobacco cell walls and change the uptake and translocation of inorganic arsenic. Frontiers in Plant Science. 2022;**13**:827453

[43] Ouyang P, Narayanan M, Shi X, Chen X, Li Z, Luo Y, et al. Integrating biochar and bacteria for sustainable remediation of metal-contaminated soils. Biochar. 2023;**5**(1):63

[44] Nayeri S, Dehghanian Z, Lajayer BA, Thomson A, Astatkie T, Price GW. CRISPR/Cas9-mediated genetically edited ornamental and aromatic plants: A promising technology in phytoremediation of heavy metals. Journal of Cleaner Production. 2023;**428**:139512

[45] Khan M, Mir JI. Potential of biotechnology in phytoremediation. Current World Environment. 2021;**16**(2):622

[46] Sorour AA, Khairy H, Zaghloul EH, Zaghloul HA. Microbe-plant interaction as a sustainable tool for mopping up heavy metal contaminated sites. BMC Microbiology. 2022;**22**(1):174

[47] Fuchs B, Saikkonen K, Damerau A, Yang B, Helander M. Herbicide residues in soil decrease microbe-mediated plant protection. Plant Biology. 2023;**25**(4):571-578

[48] Feng Y. Interactions among rhizosphere microorganisms, mechanisms and potential application in phytoremediation. In: SHS Web of Conferences. Vol. 144. France: EDP Sciences; 2022. p. 01003

[49] Priya AK, Muruganandam M, Ali SS, Kornaros M. Clean-up of heavy metals from contaminated soil by phytoremediation: A multidisciplinary and eco-friendly approach. Toxics. 2023;**11**(5):422

[50] Verma RK, Sankhla MS, Jadhav EB, Parihar K, Awasthi KK. Phytoremediation of heavy metals extracted from soil and aquatic environments: Current advances as well as emerging trends. Biointerface Research in Applied Chemistry. 2021;**12**(4):5486

[51] Zanganeh F, Heidari A, Sepehr A, Rohani A. Bioaugmentation and bioaugmentation–assisted phytoremediation of heavy metal contaminated soil by a synergistic effect of cyanobacteria inoculation, biochar, and purslane (Portulaca oleracea L.). Environmental Science and Pollution Research. 2022;**29**(4):6040-6059

[52] Bilgaiyan P, Shivhare N, Rao NG. Phytoremediation of wastewater through implemented wetland–a review. In: E3S Web of Conferences. Vol. 405. France: EDP Sciences; 2023. p. 04026

[53] Arivukkarasu D, Sathyanathan R. Phytoremediation of domestic sewage using a floating wetland and assessing the pollutant removal effectiveness of four terrestrial plant species. H2Open Journal. 2023;**6**(2):173-187

[54] Taufikurahman T, Melani L, Sianturi NLH, Wirawan NKDA, Saifullah AF. Effective removal of chromium from contaminated water using horizontal subsurface flow constructed wetland with Napier grass (Pennisetum purpureum). Current Research on Biosciences and Biotechnology. 2023;**4**(2):275-282

[55] Enitan IT, Durowoju OS, Edokpayi JN, Odiyo JO. A review of air pollution mitigation approach using air pollution tolerance index (APTI) and anticipated performance index (API). Atmosphere. 2022;**13**(3):374

[56] Pettit T, Irga P, Torpy F. The evolution of botanical biofilters: Developing practical phytoremediation of air pollution for the built environment. In: 1st International Conference on Climate Resilient Built Environment iCRBE. United Kingdom: World Energy and Environment Technology Ltd-WEENTECH; 2020

[57] Rabiee M, Kaviani B, Kulus D, Eslami A. Phytoremediation potential of urban trees in mitigating air pollution in Tehran. Forests. 2024;**15**(8):1436

[58] Vasilieva TN, Galaktionova LV, Baisheva EZ, Lebedev SV. Phytoremediation potential of woody flora in urban areas. In: IOP Conference Series: Earth and Environmental Science. Bristol, United Kingdom: IOP Publishing; 2020. p. 012116

[59] Ilić ZH, Solomun MK, Šumatić N. Phytoremediation: Nature Based Solution for Contaminated Urban Soils

[60] Hostyn G, Schwartz C, Côme JM, Ouvrard S. Assessment for combined phytoremediation and biomass production on a moderately contaminated soil. Environmental Science and Pollution Research. 2022;**29**:11133-11145. DOI: 10.1007/s11356-022-19963-9

[61] Fahad S, Chavan SB, Chichaghare AR, Uthappa AR, Kumar M, Kakade V, et al. Agroforestry systems for soil health improvement and maintenance. Sustainability. 2022;**14**(22):14877

[62] Hameed A, Hussain S, Rasheed A, Ahmed MZ, Abbas S. Exploring the potentials of halophytes in addressing climate change-related issues: A synthesis of their biological, environmental, and socioeconomic aspects. WORLD. 2024;**5**(1):36-57

[63] Choudhary A, Rijhwani S. Microbial diversity in selected agroforestry systems of Central Rajasthan. The International Journal of Life Science and Pharma Research. 2020;**10**:65-73

[64] Govindaraju M, Fowmitha Banu J, Senthamil Selvi S, Goel M. CO_2 sequestration through phytoremediation techniques with special emphasis on urban forestry to mitigate climate change impact. In: Climate Change and Green Chemistry of CO_2 Sequestration. Singapore: Springer Verlag; 2021. pp. 263-271

[65] Yasin G, Nawaz MF, Zubair M, Azhar MF, Mohsin Gilani M, Ashraf MN, et al. Role of traditional agroforestry systems in climate change mitigation through carbon sequestration: An investigation from the semi-arid region of Pakistan. Land. 2023;**12**(2):513

[66] Chukwumati J, Ikiriko M. Effectiveness of bio-phytoremediation on heavy metal contaminated soils: A review. GSC Advanced Research and Reviews. 2023;**14**(3):054-063

[67] Pang YL, Quek YY, Lim S, Shuit SH. Review on phytoremediation potential of

floating aquatic plants for heavy metals: A promising approach. Sustainability. 2023;**15**(2):1290

[68] Ibrahim R, Zubair A, Selintung M, Dwirachma AR. Phytoremediation of Fe and Mn metal in acid mine drainage using Typha Angustifolia. In: IOP Conference Series: Materials Science and Engineering Vol. 875, No. 1. Bristol, England: IOP Publishing; 2020. p. 012025

[69] Ali H, Khan E, Sajad MA. Phytoremediation of heavy metals—Concepts and applications. Chemosphere. 2013;**91**(7):869-881

[70] Umer S, Abbas Z, Aziz I, Hanif M, Abideen Z, Mansoor S, et al. Potential of ornamental trees to remediate trace metal contaminated soils for environmental safety and urban green space development. Sustainability. 2023;**15**(11):8963

[71] Mahendrakumar, Patel H, Verma SR, Solanki HA. Vertical gardens As a restoration component In urban spaces: A review. In: Sandhya R, Solanki Hitesh A, editors. Vertical Gardens as a Restoration Component in Urban Spaces: A Review (March 11, 2022). IJCRT.ORG, New Delhi, India; 2022

[72] Ogbeide O, Henry B. Addressing heavy metal pollution in Nigeria: Evaluating policies, assessing impacts, and enhancing remediation strategies. Journal of Applied Sciences and Environmental Management. 2024;**28**(4):1007-1051

[73] Kristanti RA, Tirtalistyani R, Tang YY, Thao NTT, Kasongo J, Wijayanti Y. Phytoremediation mechanism for emerging pollutants: A review. Tropical Aquatic and Soil Pollution. 2023;**3**(1):88-108

[74] Liu CJ, Deng SG, Hu CY, Gao P, Khan E, Yu CP, et al. Applications of bioremediation and phytoremediation in contaminated soils and waters: CREST publications during 2018-2022. Critical Reviews in Environmental Science and Technology. 2023;**53**(6):723-732

[75] Milla-Moreno E, Guy RD. Likelihood of using phytoremediation for mine-tailing management in Chile. International Journal of Phytoremediation. 2024;**26**(6):862-872

[76] Singh H, Pant G, Sharma J. Phytoremediation of contaminated soils: Challenges and future directions. Environmental Technology and Innovation. 2021;**23**:101606

[77] Khodaverdiloo H, Han F, Moradi N. Potentially toxic element contamination of arid and semi-arid soils and its phytoremediation. International Journal of Phytoremediation. 2020;**22**(5):432-444

[78] Sahoo S et al. CRISPR/Cas-mediated functional gene editing for improvement in bioremediation: An emerging strategy. In: Omics Insights in Environmental Bioremediation. Singapore: Springer Nature Singapore; 2022. pp. 635-664

[79] Cao J, Tan Y, Zhang C. Combined approaches for the remediation of cadmium- and arsenic-contaminated soil: Phytoremediation and stabilization strategies. Applied Sciences. 2024;**14**(16):7144

[80] Mocek-Płóciniak A, Mencel J, Zakrzewski W, Roszkowski S. Phytoremediation as an effective remedy for removing trace elements from ecosystems. Plants. 2023;**12**(8):1653

Chapter 2

Application of Fruit Wastes in the Bioremediation of Heavy Metals: Mechanism, Challenges and Prospects

Felicia Omolara Afolabi and Paul Musonge

Abstract

Heavy metals are toxic substances which are hazardous to plant, animal and human health. The presence of these pollutants in the environment can have adverse effects that can threaten human health if inhaled or absorbed. The bioremediation of heavy metals using fruit wastes offers the application of agricultural wastes and low-cost biomaterial and is eco-friendly for the treatment of wastewater. Fruit peels are promising and sustainable waste material for the effective treatment of wastewater, which support the concept of circular economy and environmental sustainability. This chapter provides an overview of bioremediation of heavy metals while highlighting the application of fruit peels as a promising waste material for effective treatment of wastewater. Given the success of bioremediation using agricultural wastes, the mechanisms involved in the biosorption process are discussed. Challenges and prospects for the efficient application of fruit wastes for the bioremediation of heavy metals are discussed.

Keywords: bioremediation, fruit peels, mechanisms, heavy metals, eco-friendly, wastewater

1. Introduction

The occurrence of heavy metals in the environment, majorly soil and water, has become challenging and alarming. This is due to consistent increase in urbanization and technological advancement. The heavy metal sources can be traced to several industrial activities such as smelting, galvanizing, metal plating, paint, textile, laundry processes, thermal power plant and so on. These pollution sources are increasing with time and producing huge quantity of pollutants to water bodies, thus reducing the quality and accessibility to fresh and clean water. It is stated that the major sources of water pollution are industrial discharges, agricultural activities and municipal waste discharges. There are rules and regulations by government and protection agencies to minimize the discharge of industrial effluent and water pollution. However, the industries are faced with high cost of treatment processes due to the large flux of water generated, thus narrowing the profit margin of the industries. Hence, there is need for sustainable and economical wastewater treatment techniques.

Basically, the wastewater treatment process can be divided into three phases, namely primary, secondary and tertiary treatment. These phases involve several techniques employed in water engineering such as biological, chemical, coagulation/flocculation, ion-exchange, precipitation, electrochemical, adsorption and other combinations processes. Adsorption process is a renowned wastewater treatment technique for the removal of heavy metals and other organic pollutants. Adsorption is well applied for treatment of wastewater due to its versatility, ease of operation, and economics. The efficiency of the adsorption process depends on the choice of adsorbent. A suitable adsorbent must possess high affinity for the selected pollutant to achieve good percentage removal and efficiency. The sources of these adsorbents could be natural or synthetic. Some of the pronounced adsorbents include activated carbons (ACs), metal organic frameworks, zeolites, biomass, carbon nanotubes, graphene-based materials, polymeric materials and so on. Recently, researchers have focused on the application of naturally occurring materials such as agricultural wastes as adsorbents for wastewater remediation. These biological materials are abundant in nature, readily accessible, inexpensive and sustainable. These biowastes are mostly trashed in the bin and eventually end up in the dumping site or left to rot on the farmland. In the past decade, researchers have proven that these biomaterials contain active sites suitable for the adsorption of heavy metals and organic matter. Fruit wastes have high lignocellulosic composition comprising of cellulose, hemicellulose and lignin. These compositions contain chemical functional groups responsible for the adsorptive potential of the bio-adsorbents. However, the performance and behavior of these bio-adsorbents during the adsorption process depends on their preparation, composition and different modification techniques.

Fruit wastes are a major contributor to biomass wastes. These wastes include orange peel, banana peel, watermelon rinds, lemon peel, apple seeds and pomace, mango peel, jackfruit seed kernel, pomegranate peel, papaya peel and seeds, grapefruit peel, coconut husks, date seeds, date seeds, pumpkin peel, passion fruit peel, custard apple seed and peel, almond wastes, tamarind seeds and husk, granadilla peels, shaddock seeds and peel, citrus limetta peel and so on [1–15]. Different adsorbents derived from fruit wastes have been explored for the treatment of wastewater. The fruit-based bio-adsorbents have been investigated for the removal of heavy metals and other contaminants [1, 16]. A family of adsorbents called "fruit waste derived bio-adsorbents" offers a more comprehensive view of their uses in the treatment of wastewater. These adsorbents are highly appealing, and feasibility studies indicate that using these bio-adsorbents for large-scale applications will be profitable.

This chapter elaborates different sources and preparation of bio-adsorbents derived from fruit wastes for the removal of heavy metals from wastewater. The modification techniques and biosorption capacities of the fruit wastes and adsorption isotherm and kinetics of the various fruit wastes of bio-adsorbents are discussed. The effects of the operating parameters such as initial concentration, adsorbent dosage, pH, contact time and the adsorbent particle size on the adsorption process are discussed. Furthermore, the adsorption mechanism and the challenges with the application of fruit wastes in adsorption studies are also provided.

2. Sources and preparation

Fruit wastes are naturally and abundantly available biomass which are a rich polymer source. Natural fruit is a type of cellulose-rich biopolymer source, and as **Figure 1** illustrates, their wastes are separated into various sections. These wastes have a high mechanical strength functionality and a high chemical surface reform due to their hydrophilic character. The availability, preparation and modification of the fruit waste bio-adsorbents are discussed in this section.

2.1 Sources and availability

The need for and consumption of food and fruit has increased to billions of tons annually due to population growth. This suggests that a lot of fruit wastes, like peels, leaves, stalks, seeds, stones and so forth, are produced. These wastes can be collected from the marketplace, household bins, restaurants and fruit juice industries. High quantities of these wastes end up in the dumping site, which pollute the environment or become nuisance in the community if not well disposed of. **Table 1** lists the quantities of these fruit wastes and shows that a significant amount of specific fruit waste is thrown away annually. It has also been noted that excessive garbage production might pollute the environment and increase waste management expenses. As **Table 1** illustrates, the fruit wastes have a high lignocellulosic component. They contain around 50% carbon elements, which can be transformed into activated carbons by high-temperature thermal treatment procedures carried out in an inert atmosphere. As of right now, fruit waste bio-adsorbents have been shown to be effective at removing pollutants from wastewater. Fruit waste can be used as biochar (BAs) for wastewater treatment. Examples of these are bananas, oranges, watermelon, pomegranates, strawberries, pineapples, avocados, jackfruit, mangos, coconuts, apricots, peaches, walnuts, peanuts, dates, cherries and almonds.

2.2 Preparation

The preparation of bio-adsorbents derived from fruit wastes is an easy and simple process. The bio-adsorbents can be utilized in its natural form or be subjected to

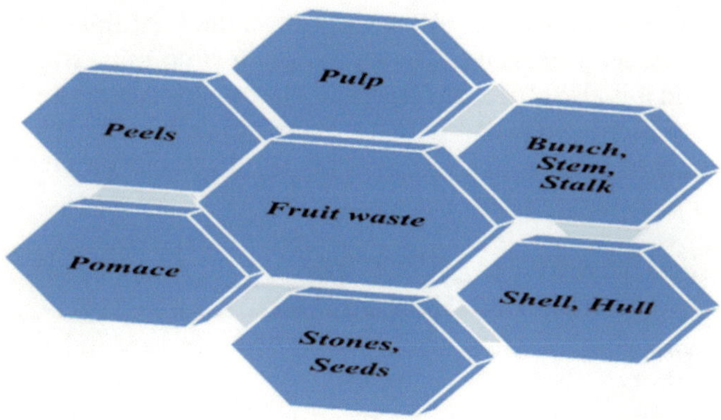

Figure 1.
Different parts of fruit wastes.

Fruit	Waste	Lignocellulosic composition (dry basis)			Waste volume MTY	Reference
		Cellulose	Hemicellulose	Lignin		
Banana	Peel	9.9	41.38	8.9	41.265	[17]
	Leaves	35.58	23.46	10.58	320,000–400,000 T	[16]
	Stem	38.48	25.36	5.77	1000	[17]
Orange	Peel	24.0	50.8	20.7	25–30	[18]
Watermelon	Peel	58	14	11	42	[19]
Pomegranate	Peel	7.9	8.1	22.1	1.5	[20]
Lemon	Peel	23.1	8.09	7.6	57.2	[21]
Citrus	Peel	20.8	17.2	8.9	52.2–57.2	[21]
Pineapple	Skin	20.44	9.43	41.21	0.62	[22]
Avocado	Peel	12.1–27.6	11.5–25.3	4.4–35.3	1,050,000	[23]
	Stone	6.5–40.9	3.0–47.9	1.8–15.8	1,500,000	[23]
Tamarind	Husk	14.85	33.40	25.77	0.17	[24]
	Seeds	20.93	15.77	28.76	0.055	[24]
Jackfruit	Leaves	34	4	19	—	[25]
Mango	Leaves	40	11	28	—	[8]
	Peels	38.35	13.90	27.90	35	[26]
	Stone	49.99	21.15	25.53	60	[26]
Coconut	Shell	48.9	19.8	30.1	2.31 T	[27]
Peanut	Shell	31.5	18.8	50	11,000,000 T	[28]
Dates	Seed	23.9	26.8	21.6	900,000 T	[29]

Table 1.
Waste volume of fruit wastes and their lignocellulosic composition.

further treatments and modification to enhance the properties and the active sites. The preparation processes can be grouped into three stages (**Figure 2**).

Most of the studies chose to use bio-adsorbents in their natural form without any modification or further processing. This simplifies the processing stages and saves time and cost, and it is also efficient. Generally, the bio-adsorbents are firstly sourced or collected and washed with water to remove dirt and debris. The type of water used for washing ranges from tap water, distilled water and deionized water to double

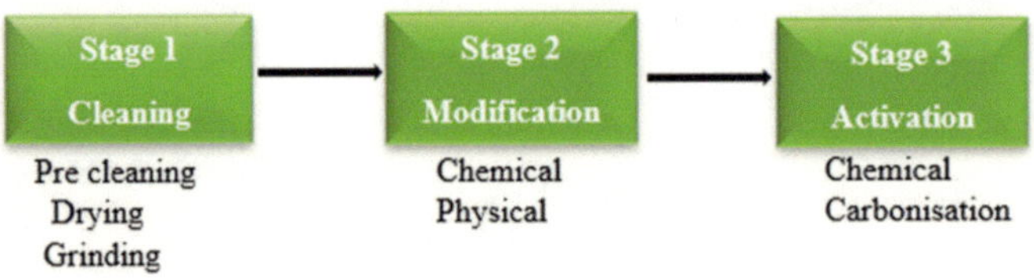

Figure 2.
Preparation stages of bio-adsorbents.

distilled water. Washing is mostly done to get rid of any dirt, dust or other particles that have adhered to the leaves' surface. After the powder has been made, washing is occasionally carried out to get rid of soluble materials, colors, tannins and other things.

Reducing moisture through drying is the second important stage in the production of bio-adsorbents from fruit wastes. There are three major techniques employed by researchers for effective drying of bio-adsorbents, namely.

1. Sun drying

2. Ambient temperature drying

3. Oven drying

4. Combination of the above 3 techniques

While some research only utilized sun drying as a first stage of drying, others reported using it exclusively. Sun drying requires a lot more time than oven drying; in some cases, it might take up to 7 days. Most of the research, however, employed oven drying. There was a range of 50–110°C in temperature and 2 hours to 2 days in duration [30–32]. Drying was also ended when crisp without giving a precise time. Room-temperature air drying was used by certain studies. Grinding and sieving are the third important step in the preparation process. The bio-adsorbents are sieved to particle sizes using a range of mesh diameters after being ground. To enhance their functional qualities, the bio-adsorbents can be utilized either exactly as it is or through additional changes.

2.3 Modification

The standard procedure for altering bio-adsorbents involves soaking them in an acid or an alkali for a few hours while shaking them occasionally at high temperatures. Formic acid, hydrochloric acid, sulfuric acid and nitric acid are among the acids that are employed. The alkali that is most widely used is sodium hydroxide. Other compounds utilized for alteration include formaldehyde, iron oxide, calcium oxide, sodium sulphide, methanol, detergents and many others. Additionally, reports of extremely complex multi-stage pre-treatment procedures can be found in the public literature.

In other investigations, the materials underwent chemical activation followed by carbonization in place of chemical modification. Compounds including zinc chloride, sulfuric acid and phosphoric acid are used in activation. The typical range for carbonization temperature was 400–1200°C; some experiments used microwave heating with a specific heating power. The time range for carbonization was 1–24 hours. For several reasons, washing is typically performed following chemical pre-treatment.

1. To eliminate extra reagents from the bio-adsorbents following the modification procedure.

2. To make sure that, even after more washing, the pH is neutral or no longer changing.

3. Biosorption capacity of fruit peels

The removal efficiency and adsorption capacity of a material are important indicators that help assess its effectiveness as an adsorbent in eliminating a particular pollutant from an aqueous solution. While the removal effectiveness is dependent on various circumstances such as the concentration of the pollutant, the existence of competing chemical species in the aqueous phase and time, the adsorbent's inherent attribute towards the pollutant is its adsorption capacity. It should be mentioned that the removal efficiencies were occasionally shown as values that were more than (>) or less than (<). These are the values that were determined by visually examining the graphs in the articles where precise values were not stated. The removal efficiencies and adsorption capacities given in the articles are maximum values (at the optimal factor levels), not averages.

3.1 Effects of operating parameters

In relation to the biosorption of contaminants from aqueous solution by fruit wastes, a few aspects have been investigated. They consist of process temperature, adsorbent dosage, pH, and equilibration time. We will look at these factors' optimums and effects in this section. Temperature has a minor impact on biosorption at medium temperatures (20–35°C) and a significant impact at higher temperatures. Elevated temperatures have the potential to significantly denature the bio-adsorbent. However, after the biomaterial has been carbonized, this is not considered. According to Anwar et al. [33], pH is the most crucial factor in the biosorption process. It is possible to think of biomass as naturally occurring ion-exchange materials, the majority of which are basic and mildly acidic. As a result, the pH of the solution impacts the metals' solution chemistry, the activity of the functional groups in the biomass and the competition between metallic ions, as well as the type of the binding sites in plant leaves and their solubility in metals. The ideal biosorption period varies considerably throughout research. This depends on the kinetics of the biosorption process to a large degree. The ideal dosage is very individualized and will never be the same for two studies. On the other hand, an increase in dosage generally results in an increase in the amount of heavy metals adsorbed. This makes sense because a higher dosage translates into more active sites that are available for adsorption.

3.1.1 Adsorbent dosage

One of the key components in creating an effective adsorption system is optimizing the adsorbent-dose ratio. It aids in estimating the treatment cost of an adsorbent and establishing the equilibrium relationship between the adsorbent and the adsorbate. The availability of more empty sites on the adsorbent for the adsorbate in solution is what causes the rise in adsorption capacity at higher amounts of adsorbent. Furthermore, when the dose of the adsorbent is changed, its surface area is likewise increased. Moreover, interactions between adsorbed ions and ions in solution might occasionally cause a larger dose of adsorbent to cause the desorption of adsorbed ions from the adsorbent's surface. As a result, for pollutants to be removed effectively, the adsorbent dose must be tuned. Numerous investigations on metal adsorption have been conducted utilizing various natural adsorbent materials. In most of the results reported, the adsorption capacity increased as the adsorbent dosage increased [5].

3.1.2 pH

The adsorption system is dependent on the pH of the solution due to the variation in the surface characteristics of the adsorbent and adsorbate at different pH levels. The removal percentage of phenol was shown to decrease as solution pH increased from 1 to 8 when phenol adsorption was carried out using chemically activated banana peel adsorbents. This was due to the larger concentration of OH- ions at higher pH levels. The disruption in phenol ion transport caused by an increase in hydroxyl ion concentration in solution lowers the adsorption capacity. As pH rises, the oxidation state of chromium changes from the more stable Cr(III) to the more hazardous Cr(IV). When it comes to lead, lower pH levels result in higher concentrations of free lead, while higher pH values cause immobilization due to precipitation [34]. When it comes to metal adsorption, rising pH causes the adsorbent surface to become negatively charged, which makes it easier for positively charged heavy metal ions to sorb [35].

3.1.3 Particle size

One special way to change an adsorbent's sorption capacity is through physical alteration. Smaller particle size adsorbents hold a larger adsorbent surface area and higher active unoccupied sites that lead to better sorption capacity. The available adsorbent surface area for adsorbate is estimated by the particle size distribution. The widening of holes and pores on the surface of adsorbents is contributed by a higher surface area. In the adsorption study of Cu^{2+} and Pb^{2+} using orange peels [36], the effect of particle size was investigated in the range of 75–455 microns. The increase in the particle size had little or no effect on the adsorption capacity of the metal ions hence 75 microns gave the optimum adsorption efficiency.

3.1.4 Contact time

An adsorbent's sorption capacity rises as the contact/interaction time does. Dyes' absorption first rises with an increase in contact duration before reaching an optimal value. Because there are many open adsorbent sites at the beginning of the adsorption process, maximum sorption occurs there. Eventually, the rate of adsorption reduces, creating a straight line on the adsorption amount against time that indicates the adsorbent's saturation level. The best time to choose for the adsorption process is crucial since it reduces operational costs and time. The amount of pollutant sorption reduces when the adsorbent and solution have more contact time. In a study conducted to investigate the effect of contact time in the adsorption of Cd^{2+} using banana peels [37], the increased adsorption of cadmium was observed with the increasing of contact time. The result showed that maximum adsorption (85% removal) of cadmium occurred at the first 10 minutes of contact time. Adsorption rate gradually increased up to 50 minutes of contact time. After 50 minutes, less increase in adsorption rate was observed and remained almost constant after 120 minutes. Optimum contact time for cadmium adsorption was selected as 120 minutes for further experiments.

3.1.5 Temperature

One of the key process variables that establishes the endothermic or exothermic character of an adsorption process is temperature. Adsorption treatment procedures are often endothermic. The bio-adsorbent sorption capability increases with temperature. Pollutant molecules' mobility or the expansion of active sites on the adsorbent's surface could be the cause of the increase in sorption capacity. Higher temperatures cause the sorbent's interior structure to enlarge, which allows dye molecules to penetrate even further [38].

3.1.6 Initial concentration

The higher initial concentration of pollutant in the wastewater results in higher removal of the pollutant and vice versa. The orange peels and magnetized orange peels have been used as adsorbents for the removal of crystal violet dyes from the textile wastewater. In a study on the adsorption of Cu^{2+} and Pb^{2+} onto sawdust [39], an increase in the Pb concentration from 1 to 50 mg/L results in an increase in the percent metal removal for up to 25 mg/L Pb for contact times above 3 hours and up to 10 mg/L Pb for lower contact times. The decrease in metal removal percentages, whenever observed, can be related to saturation of available binding sites on the adsorbent above a certain concentration of metal. The increase in adsorption capacity may be due to the higher adsorption rate and the utilization of all available active sites for adsorption at elevated metal concentration.

3.2 Isotherm studies

The adsorbate's adsorption uptake behavior on the adsorbent surface is described by adsorption isotherms. This sheds light on the forces of attraction and the affinity between the sorbent surface and the sorbate. However, it offers a necessary situation to produce monolayers and multilayers on the sorbent surface. Furthermore, sorbate adsorption on the adsorbent surface might result in the creation of a single molecular layer (monolayer formation) or, if sorbate adsorption is permitted on the surface, the accumulation of many layers (multilayer). To investigate the adsorption mechanisms, the Langmuir and Freundlich isotherm models are most utilized. At the molecular level, each of the isotherm models has a distinct sorption mechanism. These models are adapted to experimental data under isothermal conditions.

One of the isotherm models that is frequently used to characterize and identify monolayer adsorption systems is the Langmuir model. The link between the free adsorbate molecules/vacant sites on the adsorbent materials and the unavailable site on the adsorbent material is known to be inferred from this equilibrium [40]. The maximal monolayer adsorption on the adsorbent is predicted by the Langmuir adsorption theoretical concept. The distribution of metal ion equilibrium between the liquid and solid stages is correlated with the creation of a single monolayer on the adsorbent outer surface. There are no interactions between lateral adsorbed molecules, uniform adsorption sites across the surface, and finite adsorption sites in the case of monolayer adsorption. There are only a limited number of adsorption sites on the monolayer surface, which have a constant adsorption energy and no adsorbate transmigration. In summary, adsorption becomes clogged due to saturation of the adsorption sites once they have been filled on a homogenous surface and do not interact with any more adsorbed species.

Using a semiempirical formulation for the adsorption, the Freundlich model is used for adsorption uptake, which arises from both multilayer and heterogeneous adsorption layers. The basic premise of Freundlich isotherm model indicates that monolayer adsorption facilitates the adsorption on heterogeneous surface by the result of wastewater treatment activity [41]. The Freundlich model postulates that there is a non-stable, continuously stepping ratio between the solute concentration in the solution and the solute uptake quantity onto the adsorbent mass. This is the reason that the adsorption on the heterogeneous surface and multilayer adsorption are screened by the Freundlich equation.

3.3 Kinetic studies

One of the key components of adsorbents is the adsorption kinetics data. These empirical or semi-empirical models are essential for assessing and verifying the experimental results, together with isotherms. The Weber-Morris intraparticle diffusion (WMIPD), Elovich eq. (EL), Bangham equation (BNG), intraparticle diffusion (IPD), pseudo second order (PSO), pseudo first order (PFO), and so on are a few of the important kinetic models. In 1898, Lagergren initially introduced the pseudo first order kinetics model, which has been used to depict the sorption process in many research studies. One way to put it is that the number of accessible adsorption sites and the rate of sorbate ion uptake are related. According to the pseudo first order kinetics model, the adsorption rate appropriately depends on the expression of concentration differences rather than the bulk solute concentration in the solution. This is because the adsorption rate is relatively related to the adsorption capacity available during the reaction time.

4. Mechanism of adsorption

Numerous processes exist for the bio-adsorbent to absorb heavy metals; varying theories have been proposed in the literature, depending on the bio-adsorbent type and categorization standards. Furthermore, because the biological nature of the bio-adsorbents involves complexations, most mechanisms are not fully known in detail. The mechanisms are either characterized according to (a) the dependence on cell metabolism or (b) the location within the cell where the metal is eliminated, which is separated into three types: extra cellular accumulation/precipitation, cell surface sorption/precipitation and intracellular accumulation.

Bioaccumulation has replaced the term "biosorption" as the term used to describe the biosorption based on cellular metabolism. Transport across the cell membrane and precipitation are the mechanisms of biosorption or bioaccumulation that are dependent on cell metabolism, whereas precipitation, physical adsorption, ion exchange and complexation are the mechanisms of biosorption that are independent of cell metabolism. Several variables, such as the metal of interest's chemical, stereochemical and coordination properties—which include the metal ion's mass, ionic radius, and oxidation state—can influence the mechanisms. The number of reactive binding sites, the type of binding site, the accessibility and availability of binding sites and the affinity between the binding site of the bio-adsorbent and the metal ion of interest are some of the parameters of the bio-adsorbent of choice that influence the biosorption mechanism. In some circumstances, more than one mechanism may take place in a multi-step process mechanism; biosorption of uranium by Aloe vera waste includes physical adsorption, ion exchange, complexation and precipitation [42].

4.1 Physical adsorption

When working with microorganisms, physical adsorption occurs on the surface of the bio-adsorbent with the help of electrostatic interactions like Van der Waals forces and may even involve the cell walls. The bio-adsorbent's surface area and, occasionally, the pH of the solution have an impact on the physical adsorption mechanism. Physical adsorption was only favored under basic conditions in the biosorption of lead by Hami melon peels, with the biomass's carboxylate and hydroxylate groups contributing [43].

4.2 Ion exchange

Ion exchange process is the replacement of an ion bounded into the solid phase, which is readily exchangeable with another ion in solution [41]. Furthermore, it was discovered that certain biosorption processes, such ion exchange, might be impacted by the pH of the solution. At low pH levels, the biosorption mechanism of lead ions by Hami melon peels was accomplished through ion exchange; at higher pH levels, the process was accomplished through electrostatic interactions between the lead ions and the carboxylate and hydroxylate groups on the biomass [43].

4.3 Complexation

The advanced form of complexation is chelation, in which the organic ligand would bond coordinatively with the metal ion from more than one position at the same time to form a complex of higher stability. Complexation is the process of complex formation by electrostatic attraction or covalent bonds between metal ions and organic molecules acting as ligands with the ability to donate electrons. The Hard and Soft Acids and Bases (HSAB) theory suggests that elements are classified into hard or soft acids (mainly metals) and hard or soft bases (mainly non-metals). Hard acids have a stronger affinity to hard bases and therefore the weak acids to weak bases. This theory underpins the affinity of metal ions and organic ligand, which is the most important factor in complexation. Since lead is a soft acid, it will bind more readily to organic ligands of the bio-adsorbent that contain soft bases like nitrogen or sulfur. Similarly, ions with borderline hard/soft characteristics like manganese, zinc, cadmium, and copper will also readily form complexes with organic compounds that contain nitrogen or sulfur. However, the availability and accessibility of the binding site that houses the donor atom (base) have an impact on complexation.

4.4 Precipitation

Precipitation refers to the development of insoluble metal in the form of precipitate, which is one of the few mechanisms involved in the metabolism-dependent biosorption, although metabolism-independent biosorption can also occur by precipitation. Precipitation is the result of the microorganisms' active defense mechanism reacting to an environment containing harmful metal ions in metabolism-dependent biosorption. Precipitation results from chemical reactions between the metal ion and the functional groups of the bio-adsorbent's cell wall in metabolism-independent biosorption; these reactions may involve oxidation-reduction processes.

5. Reusability and disposal of bio-waste adsorbent

Reusing adsorbent helps ensure process economic viability, which allows for the appropriate exploitation of an adsorbent's potential. Fruit-based bio-adsorbents' reusability lowers wastewater treatment costs while simultaneously strengthening the lengthy cycle process. The re-adsorption and desorption procedures are carried out multiple times to reuse the bio-adsorbent. In one such instance, a bio-adsorbent produced from kiwi fruit and enhanced by magnetic chitosan was designed to adsorb patulin [44]. According to the study, the produced adsorbent's adsorption effectiveness peaked at 89% in the first step, while the desorption rate dropped to 67.7% following ethyl acetate regeneration. The adsorption and desorption rates then progressively dropped to 87.15 and 70.5% for adsorption and 51.3 and 34.7% for desorption, respectively, for the second and third cycles.

Additionally, it is quite beneficial to dispose of old adsorbent, which assesses the entire environmental impact. After the biomass has decomposed, the leftover adsorbents can typically be employed as soil fertilizers, which otherwise could pollute the environment. In this context, numerous techniques for getting rid of spent adsorbents have been developed [45]. For instance, at 473 K, fluoride-loaded adsorbents are transformed into ash by leaching fluoride, and the fluoride-free ash is then utilized for brickmaking or road construction [46].

6. Challenges with fruit wastes and prospects

After being chemically or physically activated, fruit waste can be used as a sustainable alternative to traditional, expensive commercial adsorbents for the management of solid waste. The use of hemicellulose and lignin as adsorbents has garnered a lot of attention lately due to their increased concentration and relative affordability. Fruit wastes are further investigated in terms of their properties, such as porosity, specific surface area, adsorption rate towards contaminants and so on, prior to being used as adsorbents. The expense of chemicals needed for their synthesis and regeneration is one of the main obstacles in the development of bio-adsorbents obtained from fruit waste. However, to create bio-adsorbents with strong physio-chemical properties and an exceptional adsorption capacity for industrial use, environmentally sound and sustainable ways of modifying bio-adsorbents obtained from fruit waste are needed. During the off-season, fruits are more expensive. The cost of raw materials will therefore change at off-peak times. These fruit waste-based BAs, however, can be appropriately preserved for extended harvest seasons. It would be beneficial to investigate the energy-intensive technique of rapidly drying bio-adsorbents.

Another problem for waste fruit-derived adsorbents is the selective adsorption of contaminating species, which necessitates experimental methods focused on the fabrication of homogenous and selective chemical reagents. Fruit waste-derived bio adsorbents, like traditional activated carbon materials, frequently exhibit poor chemical binding that results in structural collapse and the rejection of pollutants with low adsorption rates. To effectively interact with ionic contaminants and trap them in their inner layer through appropriate bonding, fruit waste-derived bio-adsorbents must be combined with chemical modifiers. Corrosion can occasionally result from acid activation; however, some research has shown that pollutants can be adsorbed without chemical activation, which could be a cost-effective and environmentally friendly strategy. Because chemical modifiers are expensive, it is possible to optimize the physical activation settings and forego the chemical activation. In general, the focus of research should shift to investigating alternative stages of bio-adsorbents without the need for activation or chemical reagents, with a focus on precursors derived from fruit waste to create superior materials for the development of wastewater treatment applications that will pave the way for a sustainable global economy and environment.

It is crucial to adhere to the guidelines for water treatment and recycling by adsorption technology while developing and implementing adsorption technology for wastewater treatment. According to Ref. [47], the protocols cover the following: creating an optimal bio adsorbent; cleaning and drying it; characterizing its physical, chemical and morphological properties; storing it; choosing and designing the process; conducting a technoeconomic analysis and, if the process is both technically and financially feasible, designing the plant in detail. Now, adsorption column batch techniques are being researched for their potential to remove contaminants from water. Depending on the process optimization, the design and procedure of the columns that contain the adsorbent can change. Numerous papers present the results of lab-scale research on the application of various materials and columns. Nevertheless, only a few number of research studies describe the continuous process or pilot plant size applications of BAs for wastewater treatment. The reasons for this could include the need for a long retention period, waste generated during media regeneration, decreased efficiency at greater feed concentrations and so on [48]. To facilitate large-scale commercial applications, the research on the development of BAs obtained from fruit waste needs to transition from lab size to pilot scale and process technoeconomic analysis. Technical expertise and skill sets can also be established locally, where large-scale production of fruit-derived adsorbents for commercial purposes can be achieved. High cost for large-scale applications of bio-adsorbents is a hinderance for its usage at the industrial scale. The expenses comprise: (1) Material cost, which comprises labour, raw material, synthesis, raw material and adsorbent transportation, storage, chemical and other expenditures. (2) The cost of operating the adsorption process, which comprises the cost of process materials, energy, labour, maintenance and so on. (3) The cost of land, the entire infrastructure, process machinery and equipment, installation and so on is included in the capital cost. Because it requires the use of heat or chemicals, the expense of adsorbent renewal frequently results in significant operational costs. It's also important to note that the process economics of using BAs for wastewater treatment is heavily influenced by the bio-adsorbents' potential for regeneration, efficiency and recycling. The composition, shape, surface, temperature and chemical properties of BAs have a significant impact on the bio-adsorbents' efficiency, recycling and regeneration. Nonconventional adsorbents are a more appropriate term for the adsorbents generated from biomass. In addition to being environmentally benign, bio-adsorbents offer wastewater treatment as a benefit for waste management. This indicates a viable future for bio-adsorbents, particularly those produced from fruit waste. But for large-scale enterprises, it's important to highlight their advantages and solve their problems. It is essential to choose the appropriate adsorbent for the water pollutants within the specified operating parameters prior to the application of adsorption technology utilizing bio-adsorbents formed from fruit waste. The process of bio-adsorbent regeneration for the elimination of heavy metals can be difficult and expensive and result in the in-situ disposal of the adsorbents. On the other hand, bio-adsorbents appear to be more promising for eliminating toxins and organic contaminants. It is imperative to do techno-economic evaluations and feasibility studies for the treatment of diverse wastewater sources utilizing different bio-adsorbents to support lab-scale or small-scale investigations.

Food, energy and water connectivity are three of the main components to comprehend the world ecosystem and usage of resources. Because of their close interdependence, if one of these three is disturbed, the other two will also be disturbed [21]. Fruit plantations and more clean water will eventually result from the development of adsorbents produced from fruit waste for wastewater treatment operations.

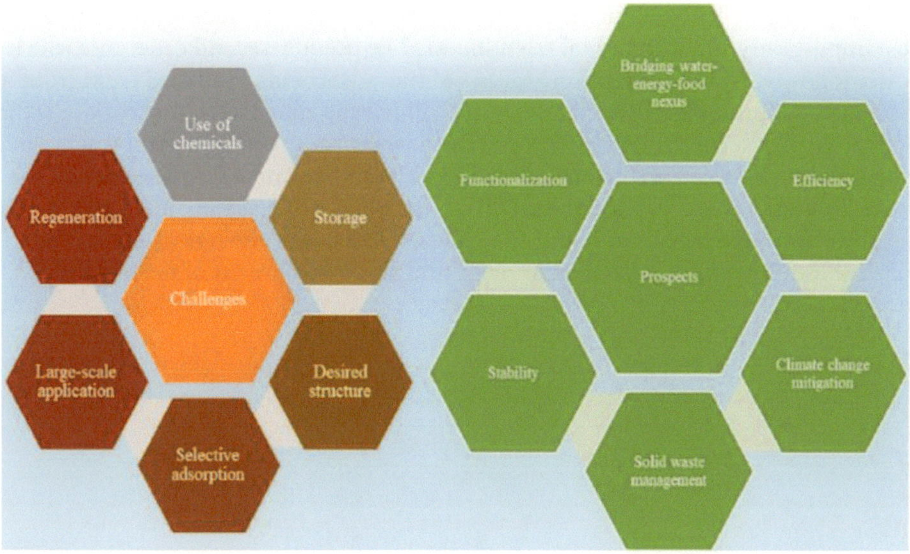

Figure 3.
Challenges and prospects of fruit wastes [1].

The development of biomass energy technology and the decrease of carbon dioxide emissions into the atmosphere are two other benefits of fruit tree plantations. The fruit tree forestation not only will improve the food-energy-water nexus but also can help in climate change mitigation by reducing carbon footprint. Since there is not a single fruit waste that can effectively address every water pollution, planting a variety of fruits locally may be a good alternative. Promoting localized fruit forests may contribute to economic viability by lowering labour and transportation costs, preventing import taxes, utilizing other available resources, developing technology, creating jobs and reducing poverty without jeopardizing food security. This calls for the coherence and integration of contemporary technologies, ranging from the harvesting, deforestation and forestry of fruit trees to the creation of bio-adsorbents and their uses in wastewater treatment and environmental remediation overall, as well as the production of biofuels. Fruit growth varies geographically throughout the world due to meteorological factors. However, a great deal of caution, policy, stakeholder acceptance and management are needed to meet the land requirements for growing food crops and optimizing the forestation of fruit trees (**Figure 3**).

7. Conclusion

Bio-adsorbents made from fruit waste have the following benefits: they are inexpensive, environmentally friendly and highly sustainable; have great adsorption; and are easy to functionalize. The potential of these bio-adsorbents to remediate wastewater is enormous. The bio-adsorbents derived from fruit waste have shown encouraging adsorption capabilities because of their wide surface area and smooth surface characteristics. Applications for bio-adsorbents and related hybrids in the adsorption of heavy metals have been investigated for waste treatment. Waste fruit-derived bio-adsorbents offer many advantages over conventional sorbents, such as carbon, zeolite, iron oxide nanomaterials and graphene-based materials. These advantages include recycling, a highly hydrophilic surface that makes it easy to surface-sorb toxins and, most importantly, environmental friendliness. The pseudo-second order rate rule is mostly followed by the adsorption kinetics of bio-adsorbents produced from fruit waste. The Langmuir isotherm model, which illustrates monolayer adsorption, is the appropriate isotherm in most situations. The adsorbents made from fruit waste can bridge the gap between water, energy and food, thereby supporting a sustainable future. Nevertheless, there are several difficulties, including scaling up, using chemicals, the effectiveness and recycling of bio-adsorbents, morphological anomalies, fruit tree localization, localized technology and the development of technical skills, among others.

Author details

Felicia Omolara Afolabi[1*] and Paul Musonge[2]

1 Department of Chemical Engineering, Durban University of Technology, Durban, South Africa

2 Institute of Systems Science, Durban University of Technology, Durban, South Africa

*Address all correspondence to: feliciaa@dut.ac.za; afolabiomolara1810@gmail.com

References

[1] Solangi NH et al. Development of fruit waste derived bio-adsorbents for wastewater treatment: A review. Journal of Hazardous Materials. 2021;**416**(125848):1-31

[2] Afolabi FO, Musonge P, Bakare BF. Adsorption of copper and lead ions in a binary system onto orange peels: Optimization, equilibrium, and kinetic study. Sustainability (Switzerland). 2022;**14**(17):1-16

[3] Aziz M et al. Efficient removal of Lead and chromium from aqueous media using selenium based Nanocomposite supported by orange Peel. Frontiers in Environmental Science. 2022;**10**:1-13

[4] Castro D et al. Chemical modification of agro-industrial waste-based bioadsorbents for enhanced removal of Zn(II) ions from aqueous solutions. Materials. 2021;**14**(9):1-17

[5] Çelebi H et al. An environmental and green process for Pb2+ pollution: An experimental research from the perspective of adsorption †. Engineering Proceedings. 2022;**19**(1):1-6

[6] David AO et al. Bio-adsorption of heavy metal ion from water using activated carbon. Kufa Journal of Engineering. 2024;**15**(2):116-127

[7] Dharsana M, Arul Jose JP. A novel green approach for lead adsorption and isotherm evaluation. Nature Environment and Pollution Technology. 2023;**22**(1):73-85

[8] Giri DD et al. Sustainable chromium recovery from wastewater using mango and jackfruit seed kernel bio-adsorbents. Frontiers in Microbiology. 2021;**12**:1-12

[9] Hawal LH, Al-Zubaidi L, Allah SA. Electro-kinetic remediation of nickel from contaminated soil using bio remedies Banana peels and surfactant-enhanced. Journal of Ecological Engineering. 2021;**22**(5):214-220

[10] Kettum W et al. Enhanced adsorptive composite foams for copper (II) removal utilising bio-renewable polyisoprene-functionalised carbon derived from coconut shell waste. Scientific Reports. 2021;**11**(1):1-12

[11] Khalid AM et al. Isolation and characterization of magnetic oil palm empty fruits bunch cellulose nanofiber composite as a bio-sorbent for Cu(II) and Cr(VI) removal. Polymers. 2021;**13**(1):1-22

[12] Mondal P, Yadav BP, Siddiqui NA. Removal of lead from drinking water by bioadsorption technique: An eco-friendly approach. Nature Environment and Pollution Technology. 2020;**19**(4):1675-1682

[13] Rangamani TP et al. Optimization and removal of heavy metals from groundwater using Moringa extracts and coconut Shell carbon powder. International Journal of Experimental Research and Review. 2023;**36**:89-98

[14] Seleman M et al. Isotherms and kinetic studies of copper removal from textile wastewater and aqueous solution using powdered banana peel waste as an adsorbent in batch adsorption systems. International Journal of Biomaterials. 2023;**2023**:1-10

[15] Wang F et al. Isotherm, kinetics, and adsorption mechanism studies of diethylenetriaminepentaacetic acid—Modified banana/pomegranate peels

as efficient adsorbents for removing Cd(II) and Ni(II) from aqueous solution. Environmental Science and Pollution Research. 2022;**29**(2):3051-3061

[16] Adeniyi AG, Ighalo JO. Biosorption of pollutants by plant leaves: An empirical review. Journal of environmental. Chemical Engineering. 2019;**7**(3):1-33

[17] Akpomie KG, Conradie J. Banana peel as a biosorbent for the decontamination of water pollutants. A review. Environmental Chemistry Letter. 2020;**18**:1085-1112

[18] Wu Z et al. Repurposing of fruit Peel waste as a green reductant for recycling of spent lithium-ion batteries. Environmental Science & Technology. 2020;**54**(15):9681-9692

[19] Dammak MI et al. Partial characterization and antitumor activity of a polysaccharide isolated from watermelon rinds. International Journal of Biological Macromolecules. 2019;**136**:632-641

[20] Saleem A et al. Acid hydrolysis optimization of pomegranate peels waste using response surface methodology for ethanol production. Biomass Conversion and Biorefinery. 2020;**12**(5):1513-1524

[21] Zhang B et al. Hydrothermal liquefaction of fresh lemon-peel: Parameter optimisation and product chemistry. Renewable Energy. 2019;**143**:512-519

[22] Lin Y-L, Zheng N-Y, Hsu C-H. Torrefaction of fruit peel waste to produce environmentally friendly biofuel. Journal of Cleaner Production. 2021;**284**:1-11

[23] García-Vargas MC, Contreras MDM, Castro E. Avocado-derived biomass as a source of bioenergy and bioproducts. Applied Sciences. 2020;**10**(22):1-29

[24] Alves JLF et al. Demonstrating the suitability of tamarind residues to bioenergy exploitation via combustion through physicochemical properties, performance indexes, and emission characteristics. Bioenergy Research. 2020;**13**(4):1308-1320

[25] Gupta M, Gupta H, Kharat DS. Adsorption of Cu(II) by low cost adsorbents and the cost analysis. Environmental Technology and Innovation. 2018;**10**:91-101

[26] Monteiro CRM et al. Hydrothermal treatment on depolymerization of hemicellulose of mango seed shell for the production of xylooligosaccharides. Carbohydrate Polymers. 2021;**253**:117274

[27] Wilson L et al. Thermal characterization of tropical biomass feedstocks. Energy Conversion and Management. 2011;**52**(1):191-198

[28] Fermanelli CS et al. Pyrolysis and copyrolysis of three lignocellulosic biomass residues from the agro-food industry: A comparative study. Waste Management. 2020;**102**:362-370

[29] Hejna A et al. On the correlation of lignocellulosic filler composition with the performance properties of poly(ε-Caprolactone) based biocomposites. Waste and Biomass Valorization. 2018;**11**(4):1467-1479

[30] Afolabi FO, Musonge P, Bakare BF. Bio-sorption of CU2+ and PB2+ ions onto *Citrus Sinensis* (Orange) Peel, in SETWM-20, ACBES-20 & EEHSS-20 Nov. 16-17, 2020, Johannesburg (SA). 2020

[31] Amin MT, Alazba AA, Shafiq M. Removal of copper and lead using

banana biochar in batch adsorption systems: Isotherms and kinetic studies. Arabian Journal for Science and Engineering. 2017;**43**(11):5711-5722

[32] Çelebi H et al. Use of banana peel for the removal of boron from aqueous solutions in the batch adsorption system. International journal of Environmental Science and Technology. 2022;**19**:1-6

[33] Anwar J et al. Removal of Pb(II) and Cd(II) from water by adsorption on peels of banana. Bioresource Technology. 2010;**101**(6):1752-1755

[34] Afolabi FO, Musonge P, Bakare BF. Evaluation of Lead (II) removal from wastewater using banana peels: Optimization study. Polish Journal of Environmental Studies. 2021;**30**(2):1-10

[35] Obike AI et al. Equilibrium and kinetic studies of Cu (II), Cd (II), Pb (II) and Fe (II) adsorption from aqueous solution using cocoa (*Theobroma cacao*) pod husk. Journal of Applied Sciences and Environmental Management. 2018;**22**(2):182-190

[36] Afolabi FO, Musonge P, Bakare BF. Application of the response surface methodology in the removal of Cu2+ and Pb2+ from aqueous solutions using orange peels. Scientific African. 2021;**13**:1-9

[37] Deshmukh PD et al. Cadmium removal from aqueous solutions using dried banana peels as an adsorbent: Kinetics and equilibrium modeling. Journal of Bioremediation & Biodegradation. 2017;**08**(03):395

[38] Romero-Cano LA et al. Functionalized adsorbents prepared from fruit peels: Equilibrium, kinetic and thermodynamic studies for copper adsorption in aqueous solution. Journal of Cleaner Production. 2017;**162**:195-204

[39] Semerjian L. Removal of heavy metals (Cu, Pb) from aqueous solutions using pine (*Pinus halepensis*) sawdust: Equilibrium, kinetic, and thermodynamic studies. Journal of Environmental Technology Innovation. 2018;**12**:91-103

[40] Taib NI et al. Kinetic, equilibrium, and thermodynamic studies of untreated watermelon peels for removal of copper(II) from aqueous solution. Desalination and Water Treatment. 2021;**227**:289-299

[41] Ali Redha A. Removal of heavy metals from aqueous media by biosorption. Arab Journal of Basic and Applied Sciences. 2020;**27**(1):183-193

[42] Noli F, Kapashi E, Kapnisti M. Biosorption of uranium and cadmium using sorbents based on Aloe vera wastes. Journal of Environmental Chemical Engineering. 2019;**7**(2):1-10

[43] Bashir A et al. Removal of heavy metal ions from aqueous system by ion-exchange and biosorption methods. Environmental Chemistry Letters. 2018;**17**(2):729-754

[44] Luo Y et al. Bioadsorption of patulin from kiwi fruit juice onto a superior magnetic chitosan. Journal of Alloys and Compounds. 2016;**667**:101-108

[45] Gomes AM, de Oliveira SM, Azevedo E, Lurling M. Temperature effect on exploitation and interference competition among *Microcystis aeruginosa, Planktothrix agardhii* and, *Cyclotella meneghiniana*. ScientificWorldJournal. 2015;**2015**:834197

[46] Mukherjee S et al. Elucidation of the sorptive uptake of fluoride by Ca 2+ –treated and untreated algal biomass of Nostoc sp. (BTA394):

Isotherm, kinetics, thermodynamics and safe disposal. Process Safety and Environmental Protection. 2017;**107**:334-345

[47] Kumar Jain C, Singh Malik D, Kumar Yadav A. Applicability of plant based biosorbents in the removal of heavy metals: A review. Environmental Processes. 2016;**3**:495-523

[48] Bhattacharjee C, Dutta S, Saxena VK. A review on biosorptive removal of dyes and heavy metals from wastewater using watermelon rind as biosorbent. Environmental Advances. 2020;**2**(100007):1-13

Chapter 3

Geochemical Processes Controlling Metals (Pb, Zn) Mobility during Amended Phytostabilization on Sulfidic Mine Tailings

Lina Xie and Dirk van Zyl

Abstract

Phytostabilization aims to stabilize metals in the rhizosphere at sulfidic mine tailings. Previous research on phytostabilization has rarely examined the geochemical processes in the rhizosphere that control acid generation and metal release. The geochemical processes that affect the mobility of Pb and Zn in the rhizosphere of phytostabilization on sulfidic tailings were explored in this chapter. The results showed that at the end of the experiments, the pH of pot substrates with plants became acidic (ranging from 2.59 ~ 3.77), and the concentrations of Pb and Zn in the drainage increased (up to 2.28 mg/L of Pb and 3199.87 mg/L of Zn). It was concluded that, in addition to the sulfide oxidation, the plants decreased pH and released metals. The addition of amendments neutralized the acidity in the rhizosphere through three geochemical processes. This research provides implications on the amendment addition during the establishment of phytostabilization on sulfidic mine tailings.

Keywords: phytoremediation, metal immobilization, acid mine drainage (AMD), inorganic geochemistry, soil acidification

1. Introduction

Sulfidic mine tailings can generate acid and release metals when exposed to oxygen and water due to the presence of sulfidic minerals [1, 2]. It is essential to reclaim and revegetate the large surface area of tailings cost-effectively. Direct revegetation can achieve this goal by amending the tailings with appropriate materials and selecting suitable plant species [3].

Phytostabilization is a promising technique for direct revegetation on mine tailings, which creates a vegetation cover to immobilize trace metals in the rhizosphere where the metals can be precipitated and stabilized [4, 5]. Previous research, through both greenhouse or field applications, has primarily focused on two key areas for enhancing plant growth and metal stabilization of phytostabilization on sulfidic mine tailings: selecting species [6, 7] and amendment materials [8, 9]. These elements are crucial for the successful establishment of phytostabilization and form the basis of the

IntechOpen

preliminary study published by Xie and van Zyl [9]. Furthermore, controlling acid generation and the release of trace metals due to sulfide oxidation is vital in this process [10, 11]. Once phytostabilization is successfully established, understanding the geochemical processes affecting metal mobility becomes critical for mitigating metal releasing (ML) and acid rock drainage (ARD) in tailings, which is the aim of this research.

The interactions between roots and tailings are valuable to assist in understanding the metal mobility of phytostabilization on sulfidic tailings by offering valuable insights into metal speciation and the migration of toxic metals [4]. In the rhizosphere, these interactions involve both organic processes, such as microbial activity and root uptake, as well as inorganic processes. Current research largely emphasizes organic processes, focusing on identifying microorganisms that enhance plant growth [12, 13], establishing microbial communities within the rhizosphere [14–16], and examining the effects of microbial activity on metal mobility [17, 18].

However, a limited amount of previous research is about the geochemistry involved in metal stabilization in the rhizosphere of tailings, as most studies have analyzed the metal concentrations and distribution rather than the underlying reactions that govern metal behavior in the rhizosphere of revegetation [19, 20]. Gaining a deeper understanding of the geochemical processes in the rhizosphere can guide the selection of more appropriate types and quantities of amendment materials [21], which increases the chance of promoting plant survival and establishing phytostabilization successfully on ARD and ML-generating tailings. Metal mobility in the rhizosphere is mainly controlled by three inorganic geochemical processes: precipitation and dissolution, silicate mineral weathering, and surface adsorption and desorption [22]. This chapter would look into these inorganic processes, specifically focusing on their role in the immobilization of lead (Pb) and zinc (Zn) of phytostabilization.

To minimize variables related to the unpredictable field conditions, the experiment was conducted in a greenhouse. The chapter aims to examine the inorganic geochemical reactions involved in Pb and Zn mobility in the rhizosphere, which fills the gap of a certain area of phytostabilization on sulfidic tailings. The results offer important insights that can be applied to practical phytostabilization efforts in real-world settings. Additionally, the results can support mining companies, governments, and stakeholders in developing more effective mine closure plans.

2. Methods and materials

2.1 Greenhouse experiment

Tailings were sourced from Faro Mine in Yukon, Canada, an abandoned lead (Pb) and zinc (Zn) mine. The selected species and amendments were previously evaluated in a preliminary study to confirm their suitability for phytostabilization [9]. In this preliminary study, wheatgrass (*Pascopyrum*) and ryegrass (*Lolium multiflorum*) were chosen for the greenhouse study, and six substrates were used as different types of growing medium in the pots. The six substrates are listed in **Table 1**. Rio Tinto provided bauxite residue, while University of British Columbia Building Operations provided compost. Lime was bought and shipped from Fisher Scientific. There were two duplicates of each pot.

Substrates	T	T + BR	T + L	T + C	T + BR + C	T + L + C
Tailings	100%	90%	100%	80%	70%	80%
Bauxite residue	-	10%	-	-	10%	-
Lime	-	-	1.25 × NAP	-	-	1.25 × NAP
Compost	-	-	-	20%	20%	20%

T = tailings; BR = bauxite residue; T = lime, C = compost; NAP = net acid potential.

Table 1.
The composition of amendments.

Each type of mixing materials was undisturbed for three days after the pots were installed, allowing particles to stabilize before sowing. During the 10-week experiment, each pot was watered to allow 100 ml drainage to be sampled weekly for analyzing pH and metal concentrations (Pb and Zn).

The greenhouse experiment lasted three months, which is the typical lifespan length of most grass species from sowing to harvesting [23], indicating that the plants would die at the end of the experiments due to aging rather than the acidity of tailings. Given that the selected species thrived on the amended tailings and developed extensive root systems that fully occupied each pot [9], it can be reasonably assumed that the bulk substrate of the amended tailings in each pot appropriately reflects the plant rhizosphere for the purpose of geochemical process analysis.

2.2 Samples analyses

The pH of rinsed substrates before and after experiments was tested by a pH meter [24]. The neutralizing acid potential (NAP) of the tailings that was analyzed by the modified Sobek Method [25] decided the amount of lime. The quantitative minerology (QXRD) of the below materials was measured: tailings, bauxite residue, and the substrates of T blank, T+BR blank, T+L blank, and T+C blank after experiments, at the University of British Columbia. After cutting the aboveground tissues at the end of the experiment, the materials in pots were oven-dried at 30°C for three days and then manually picked out roots to separately collect samples of soil and roots. Inductively coupled plasma optical emission spectroscopy (ICP-OES) analysis of Pb and Zn was conducted for all aboveground tissue, roots, and drainage samples at the University of British Columbia.

3. Results

3.1 pH of substrates

The average rinsed pH (1:5 w/w substrate: water) of materials is presented in **Table 2**, which reduced largely after the experiments. The pH in blank groups without grasses decreased to around four, and the pH in groups with grasses is even lower than four. Sulfide oxidation is one of the reasons of this acidification in the substrate with tailings, and plants caused additional acid-generating processes other than sulfide oxidation.

Substrates		T	T + BR	T + L	T + C	T + BR + C	T + L + C
Before experiments		4.69	6.09	8.43	6.55	6.22	7.97
After experiments	WG	3.38	3.15	3.22	2.59	3.77	2.88
	RG	3.72	2.71	2.90	2.33	2.97	2.35
	Blank	4.77	6.15	4.80	4.12	4.01	3.95

*T = tailings; BR = bauxite residue; T = lime; C = compost; WG = wheatgrass (*Pascopyrum*); RG = ryegrass (*Lolium multiflorum*).*

Table 2.
pH of substrates before and after experiments.

3.2 Mineralogy changes

Mineralogical alterations are the primary geochemical changes observed in the evolution of sulfide mine tailings. The percentages of mineralogical results are shown in **Table 3**. Qualitative minerology analysis was conducted for the pot samples with grass planting, and no new minerals were detected other than those listed in **Table 3**.

There are notable mineralogical changes in this experiment. The percentages of marcasite, pyrrhotite, and sphalerite reduced, and galena no longer existed, which means the process of sulfide oxidation had taken place. Additionally, the total sulfates in T Blank, T+BR Blank, T+L Blank, and T+C Blank increased from 3.5% in original tailings to 16.1% in T Blank, 10% in T+BR Blank, 11.3% in T+L Blank, and 21.8% in T +C Blank after the experiment. The sulfide oxidation led to the precipitation of sulfates such as gypsum ($CaSO_4$), anglesite ($PbSO_4$), barite ($BaSO_4$), and jarosite ($K_2Fe_6(SO_4)_4(OH)_{12}$).

The total silicates decreased from 5.5% in original tailings to 4.6% in T Blank, 4.9% in T+BR Blank, 3.5% in T+BR Blank, and 4.4% in T+C Blank after experiments. There was 10.2% sodalite, which is also a silicate mineral, in original bauxite residues. Silicate mineral weathering can be accelerated in acidic environments [26, 27], so the content of silicate minerals declined in substrates of blank groups of tailings, T + BR, T + L, and T + C after experiments.

The contents of carbonates decreased from 7.6% in original tailings to 3.6% in tailings, 7.1% in T + BR Blank, 0.9% in T+L Blank, and 0.9% in T+C Blank. Carbonates are the first type of minerals to neutralize the AMD, which contributes to the neutralization potential (NP) [28]. However, siderite ($FeCO_3$) dissolution generates Fe^{2+}, which can be oxidized to Fe^{3+}. The hydrolysis of Fe^{3+} consumes no protons, so siderite does not play a role in NP [29, 30]. The 3.1% of calcite in the original BR and the calcite generated from lime (CaO) dissolution consist of the carbonate neutralization capacity, which was consumed in the blank groups of tailings with BR and L.

Moreover, the Fe, Al hydroxides (böhmite, gibbsite, goethite) and Fe oxides (magnetite, hematite) in the blank groups of tailings, T+BR, T+L, and T+C after experiments were lower compared to those in the original bauxite residue. These indicate that the pH of drainage may drop to below four after experiment, because the dissolution of (oxy)hydroxide takes place with a pH around four in the neutralizing sequence of sulfidic tailings [31, 32].

	Mineral	Before experiments		After experiments			
		T	BR	T Blank	T + BR Blank	T + L Blank	T + C Blank
Sulfides	Pyrite	43.2%	-	46.6%	37.1%	34.5%	46.9%
	Marcasite	8.9%	-	4.1%	7.1%	4.2%	0.8%
	Sphalerite	2.3%	-	0.9%	1.2%	0.4%	0.6%
	Pyrrhotite	2.2%	-	1.4%	0.4%	-	-
	Galena	0.2%	-	-	-	-	-
Sulfates	Anglesite	0.5%	-	0.4%	0.5%	0.2%	-
	Barite	2.1%	-	8.4%	2.1%	1.2%	8.9%
	Gypsum	0.9%	-	2.5%	1.5%	2.4%	6.5%
	Jarosite	-	-	4.8%	5.9%	7.5%	6.2%
Silicates	Muscovite	2.5%	-	1.9%	2.4%	2%	2.7%
	Kaolinite	1.9%	-	0.9%	1.3%	0.8%	0.9%
	Celsian	1.1%	-	0.8%	1.2%	0.7%	0.8%
	Sodalite	-	10.2%	-	-	-	-
Carbonates	Siderite	7.6%	-	3.6%	7.1%	0.9%	0.9%
	Calcite	-	3.1%	-	-	-	-
Hydroxides	Böhmite	-	4.9%	-	-	-	-
	Gibbsite	-	1.8%	-	-	-	-
	Goethite	-	30.6%	-	0.9%	-	-
Iron oxides	Magnetite	-	-	0.9%	0.2%	0.3%	0.8%
	Hematite	-	39.4%	-	0.4%	-	-

T = tailings; BR = bauxite residue; T = lime; C = compost.

Table 3.
The percentages of quantitative mineralogy results.

3.3 pH in drainage

Figure 1(a) and **(b)** illustrates the pH of all drainage decreased regardless of the planting of grasses. In general, all pH values decreased from the first week to the experiment's end, regardless of grass species, adding amendments or not, or the amendments types, indicating that planting did not affect sulfide oxidation. The differences are their decreasing rates of various substrates, with the pH in drainage from pure tailings showing the least steep decreasing trend for both wheat and rye-grass. The pH in the first week for different amended tailings has the same order with planting either wheat or ryegrass, which is T+BR+C (purple dotted line) > T+L+C (yellow dotted line) > T+L (blue dotted line) > T+C (green dotted line) >T+BR (red dotted line) > T (black dotted line). The pH fluctuated in various ways with decreasing trends for six substrates. At the end of the experiments, the two species showed different pH values in drainage at the 10th week, and the drainage in most amended tailings had a lower pH than that in unamended tailings (black dotted line). This

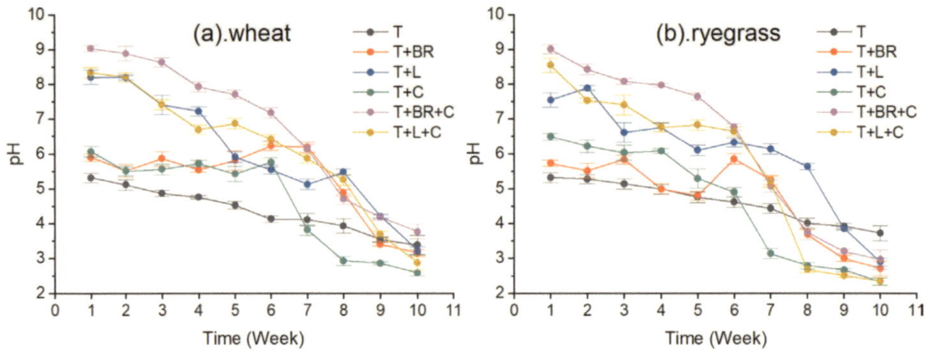

Figure 1.
pH changes of drainage from different substrates.

indicates that despite the neutralization capacity of the amendments, some processes decrease the pH of amended tailings *via* the planting of grass.

3.4 Changes of Pb and Zn in drainage

The trends of Pb concentrations in the drainage from all six substrates with or without grass during the experiments are shown in **Figure 2**. Pb concentrations in drainage from unamended tailings were at low levels when compared to those from the amended tailings. Blank groups (blue dotted lines) without grass on all substrates drained out water with low Pb concentrations. Pb concentrations started to increase from the sixth week on T + C and T + L + C with grass, and from the eighth week on T + BC, T + BR + C, and T+L with grass; these data show a more obvious increase than those of blank groups without plants. Tailings with amendments with ryegrass generated drainage with the highest Pb concentrations (red dotted lines), followed by the drainage from pots with wheatgrass (black dotted lines). In addition, concentrations of Pb varied in the drainage from tailing with different amendments.

Zn concentrations in drainage show similar features to Pb concentrations (**Figure 3**). First, Zn concentrations in drainage from unamended tailings remained at a low level when compared to those from the amended tailings. Plus, blank groups without grass on all substrates (blue dotted lines) drained water with low Zn concentrations. As Pb, Zn concentrations were at lower levels on unamended tailing and blank groups, and the unamended tailings have about the same effect on these three metal concentrations in drainage as the blank groups without plants. This is because grass can hardly survive and grow well on pure tailings due to deleterious conditions. As such, pure tailings can be considered as blank groups without grass, which is supported by the plant growth in the greenhouse experiment. Similar to Pb, Zn concentrations started to significantly increase from the sixth week on amended tailings with grass, with more obvious increases than those of blank groups without plants. Regarding the grass species, tailings with amendments with ryegrass generated drainage with the highest Zn concentrations (red dotted lines), followed by the drainage from pots with wheatgrass (black dotted lines). In addition, concentrations of Zn varied in the drainage from tailing with different amendments.

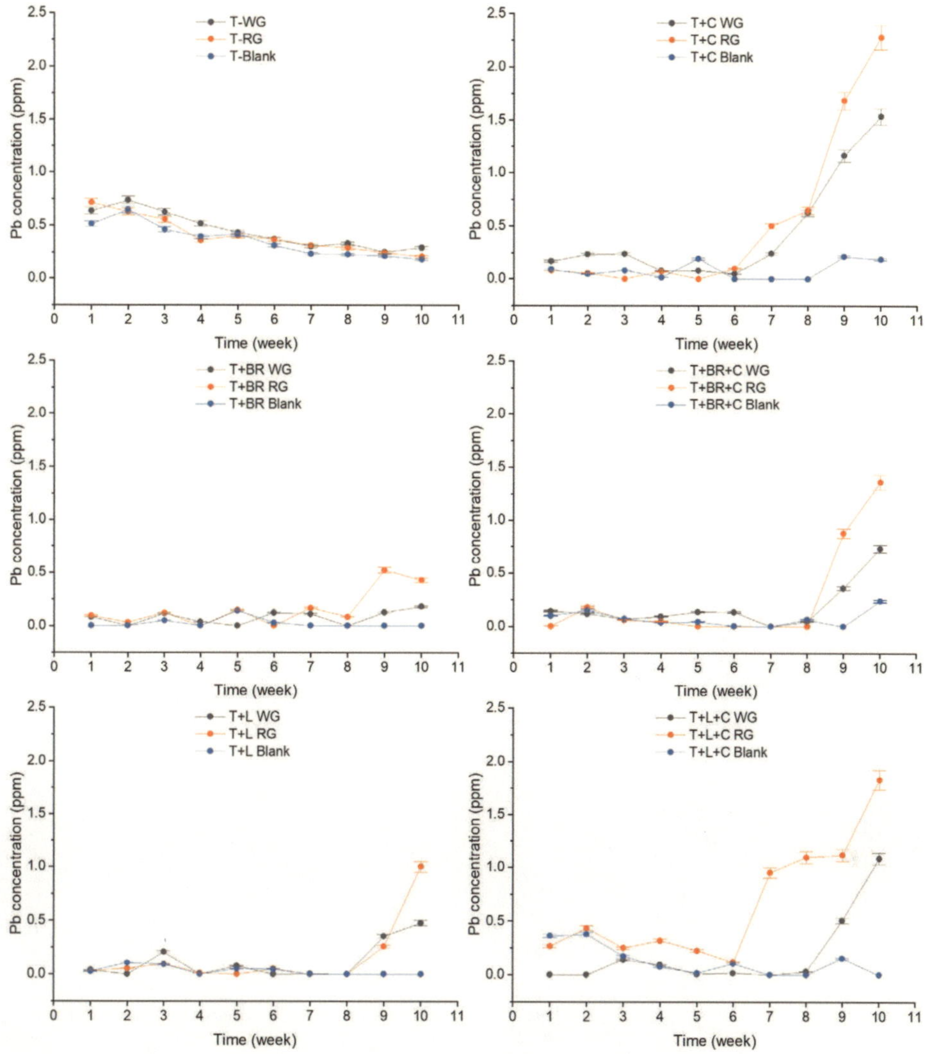

Figure 2.
Pb concentration of drainage from pots with or without plants during experiments.

4. Discussion

4.1 Acidification process

Table 3 indicates that pyrite is the most abundant sulfide mineral in the tailings sample. Pyrite oxidation has been extensively studied as it is the primary reason for the AMD formation [29, 33, 34]. In a natural environment, oxidation happens when pyrite contacts water and an oxidant (e.g., air and oxygen). The overall reaction for pyrite oxidation is as follows:

$$FeS_{2(s)} + \frac{15}{4}O_{2(aq)} + \frac{7}{2}H_2O \rightarrow Fe(OH)_{3(s)} + 2SO_4^{2-} + 4H^+ \quad (1)$$

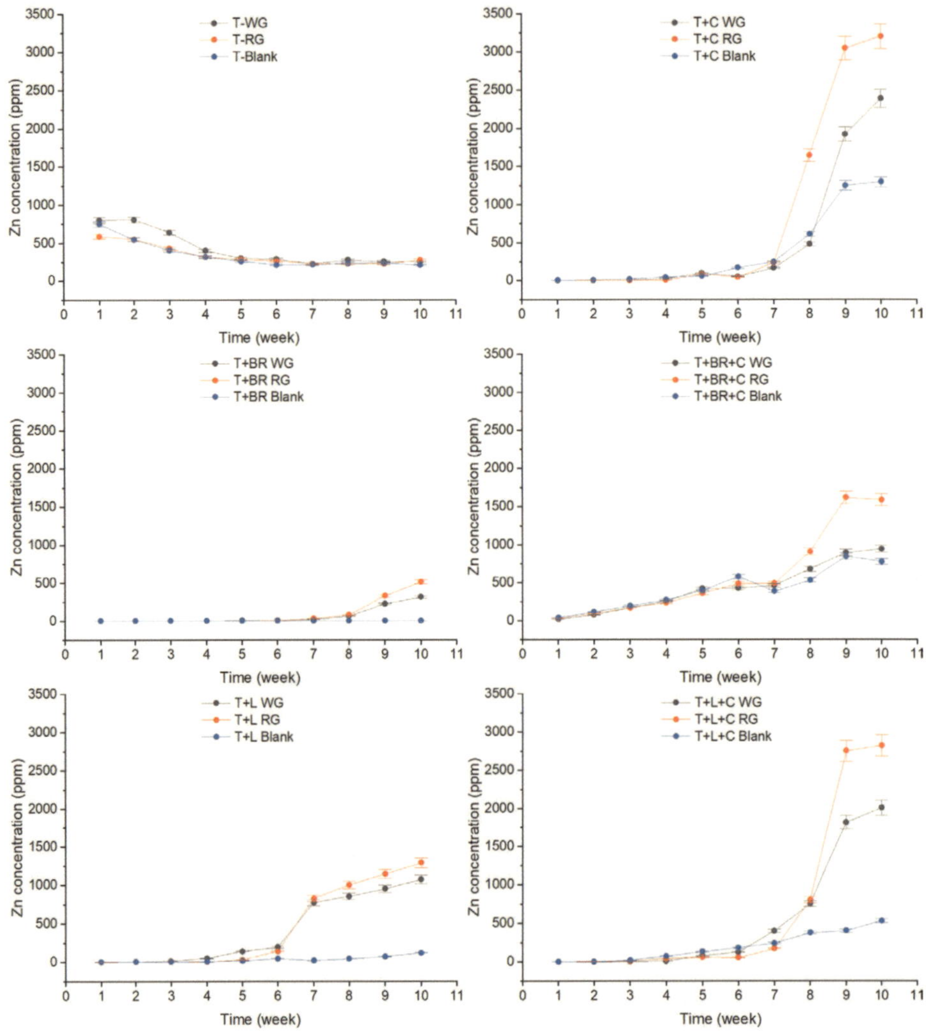

Figure 3.
Zn concentration of drainage from pots with or without plants during experiments.

The $Fe(OH)_3$ is a substitute for ferrihydrite (Fe_2O_3). Hematite (Fe_2O_3) and magnetite (Fe_3O_4) are also possible products of AMD formation [34], which is detected in pot materials after experiments (**Table 3**). Eq. (1) also proves that the ARD is generated acidic with lots of SO_4^{2-}, resulted in the increased precipitation of sulfates minerals. In this experiment, **Table 3** shows that gypsum ($CaSO_4$), anglesite ($PbSO_4$), barite ($BaSO_4$), and jarosite ($K_2Fe_6(SO_4)_4(OH)_{12}$) are those sulfate minerals generated after sulfide oxidation. Therefore, Pb comes from the oxidation of galena and Zn is from the oxidation of sphalerite. Sulfide oxidation is the major acidification process in the rhizosphere in all pots.

However, there must be other acidification processes apart from ARD generation, as the pH in drainage from groups of tailings with amendments reduced after 10 weeks even when the amendment materials are alkaline lime or bauxite residue (**Figure 2**). Soil acidification is a major environmental concern and a constraint to crop production worldwide [35]. Previous research has identified several factors that contribute to soil acidification [36, 37]:

- The precipitation of H^+ ions;
- Nutrient uptake by crops and root exudates;
- The accumulation of organic matters;
- The inputs of acidifying substances from the atmosphere;
- The application of acidifying fertilizers such as urea or ammonium salts and the planting of legumes such as clover.

In this chapter, the unamended tailings without amendments and plants represent the sole acidification process of sulfidic oxidation. **Figures 2** and **3** show that the Pb and Zn concentrations are similar in drainage from T-WG and T-RG with those from T-Blank, because T-WG and T-RG can be regarded as blank groups because of the very limited survival of wheatgrass and ryegrass on original tailings [9]. Therefore, for T Blank, regardless of the effects of grass species, sulfide oxidation is the unique reaction of acidification without other plant-causing acidifications.

For other plant-causing acidifications, the pH in drainage from tailings with amendments began to drop from the sixth week and then decreased to be lower than the pH of pure tailings in the 10th week (**Figure 1(a)** and **(b)**), particularly on T+L and T + BR where tailings were amended only with alkaline lime and bauxite residue. The pH started to drop around the sixth week, meaning alkalinity in those pots was depleted around the sixth week, resulting in the acidity being increased. As discussed above, those T groups can be regarded as blank without grasses, then the major reason contributing to acidification in T+BR and T+L is the planting. Previous research has revealed that the release of H^+ from roots results from the uptake of nutrients, primarily as cations, by plant roots, due to the electroneutrality in the soil-root system [38, 39]. No obvious differences in metal concentrations and pH were observed between the pots with wheatgrass or ryegrass, suggesting that grass species may have only a minimal influence on the degree of acidification.

Furthermore, the comparison of the pH and metal concentration results of T Blank and T+C Blank (**Figures 2** and **3**) implies that adding compost resulted in higher metal release from the T+C Blank, which indicates that compost is one reason causing the acidification. Compost is rich in organic matter, which was added to support plant growth. However, the buildup of organic matter can lead to acidification [36, 37],

Substrates	**T**	**T + BR**	**T + L**	**T + C**	**T + BR + C**	**T + L + C**
Sulfides oxidation	√	√	√	√	√	√
Compost decomposition	-	-	-	√	√	√
Roots exudates	√ (very limited)	√	√	√	√	√

T = tailings; BR = bauxite residue; T = lime; C = compost.

Table 4.
The acidification process in six substrates.

which is caused by carbonic acid (H_2CO_3) and some organic acids during decomposition [40].

In conclusion, this chapter identifies that tailings, compost, and grass species are the primary three causes of acidification, through the processes of sulfide oxidation, organic matter decomposition, and roots exudates, respectively. Sulfide oxidation is the dominant acidification reaction in tailings; sulfide oxidation and roots exudates are the main drivers of acidification in T + L and T + BR; and all three processes contribute to acidification in T + L + C, T + BR + C, and T + C, as outlined in **Table 4**.

4.2 Neutralization processes

In this chapter, lime and bauxite residue were used to adjust the pH of tailings as both are alkaline materials capable of neutralizing acidity. In addition, **Table 3** also lists other types of minerals that can accept protons to neutralize pH, which offers chances to analyze the potential neutralization processes taking place within the rhizosphere.

Previous research has revealed that the acidity in tailings is neutralized by several processes sequentially: carbonates dissolution, aluminum or iron (oxy)hydroxides dissolution, and then Al-Si bearing minerals dissolution [10, 41]. These pH buffering processes typically take place at approximately five to six, four, and three, respectively [36, 41]. Upon the completion of a neutralization process or the depletion of a specific mineral, the pH declines more rapidly. This is then followed by the onset of the next neutralization process, triggered by the presence of other buffer minerals. Finally, the pH drops to two to three when all the buffer minerals are depleted. **Figure 1** is consistent with this trend that the pH in drainage began to rapidly drop from the sixth week and reached around three at the end, indicating that all buffer minerals became depleted finally.

Carbonate dissolution is the first buffer reaction, typically the dissolution of calcite ($CaCO_3$) [28]:

$$CaCO_{3(s)} + 2H^+ \rightarrow Ca^{2+} + H_2O + CO_{2(g)} \tag{2}$$

According to the quantitative XRD results (**Table 3**), bauxite residue contains 3.1% calcite. Another material applied for amendments is lime (CaO), which neutralizes the acidity through the following reaction [28]:

$$CaO_{(s)} + 2H^+ \rightarrow Ca^{2+} + H_2O \tag{3}$$

After calcite is depleted, other carbonates such as dolomite ($CaMgCO_3$) and siderite ($FeCO_3$) continue to dissolve [42]. **Table 3** shows that the carbonate mineral in this chapter is siderite ($FeCO_3$) after the calcite depletion. However, siderite dissolution does not accept H^+ overall because of the oxidation from Fe^{2+} to Fe^{3+} and the hydrolysis of Fe^{3+} following siderite dissolution [29].

$$FeCO_{3(s)} + 2H^+ \rightarrow Fe^{2+} + H_2O + CO_{2(g)} \tag{4}$$

$$Fe^{2+} + \frac{1}{4}O_2 + H^+ \rightarrow Fe^{3+} + \frac{1}{2}H_2O \tag{5}$$

$$Fe^{3+} + 3H_2O \rightarrow Fe(OH)_{3(s)} + 3H^+ \tag{6}$$

Siderite, which was initially present in the tailings sample at 7.6%, decreased in various extent in all pots. However, siderite was not fully consumed in some pots, possibly due to the heterogeneity of the substrates.

The next buffering process is the dissolution of secondary Al-hydroxide minerals [41, 43]. Two Al-hydroxide minerals were found in bauxite residue: gibbsite (Al $(OH)_3$) and böhmite (AlO(OH)). They were consumed in T+BR+C and T+BR at the end (**Table 3**). This means that these two minerals were consumed and neutralized acidity during the experiments.

$$Al(OH)_{3(s)} + 3H^+ \rightarrow Al^{3+} + 3H_2O \tag{7}$$

$$AlO(OH)_{(s)} + 3H^+ \rightarrow Al^{3+} + 2H_2O \tag{8}$$

Sequentially, the neutralization of Fe (oxy)hydroxide is the next reaction. Goethite (FeO(OH)) is the main Fe oxyhydroxide mineral (30.6%) existing in bauxite residue in this experiment, which was dissolved and reduced to 0.9% in T+BR at the end. Meanwhile, the percentages of hematite (Fe_2O_3) in T+BR also decreased, as the minerals dissolved into Fe hydroxide and then neutralized acidity.

$$FeO(OH)_{(s)} + 3H^+ \rightarrow Fe^{3+} + 2H_2O \tag{9}$$

$$Fe_2O_{3(s)} + 3H_2O \rightarrow Fe(OH)_{3(am)} \tag{10}$$

$$Fe(OH)_{3(s)} + 3H^+ \rightarrow Fe^{3+} + 3H_2O \tag{11}$$

The next neutralization process involves the dissolution of silicates minerals at low pH levels. Although silicates dissolution can happen across a wide pH range, the rate increases under low pH conditions [41]. However, under such conditions, the dissolution of silicates is slower compared to the more soluble carbonate and Al&Fe oxyhydroxide minerals [41, 43]. The neutralization ability of silicates is especially important because most earth material types are dominated by silicate minerals, positioning them as a substantial long-term source of acid neutralization [28]. **Table 3** indicates that the silicate weathering process is exemplified by the dissolution of sodalite ($Na_4(Si_3Al_3)O_{12}Cl$) in this chapter, which existed in bauxite residue (10.2%). Like other buffering minerals, sodalite was also consumed completely in T+BR+C and T+BR after the experiment. The weathering reaction is shown in Eq. (12):

$$Na_4(Si_3Al_3)O_{12}Cl_{(s)} + 12H^+ \rightarrow 4Na^+ + 3Al^{3+} + Cl^- + 3H_4SiO_4 \tag{12}$$

Additionally, the original percentages of silicates in the tailings, including clay minerals such as muscovite ($KAl_2AlSi_3O_{10}(OH)_2$), celsian ($Ba(Al_2Si_2O_8)$), and kaolinite ($Al_2Si_2O_5(OH)_4$), decreased by the end of the experiments in the amended tailings. **Table 3** implies that the celsian dissolution resulted in barite precipitation. The neutralization reactions involved in these silicates weathering are discussed as follows [41, 44]:

$$KAl_2AlSi_3O_{10}(OH)_{2(s)} + 10H^+ \rightarrow K^+ + 3Al^{3+} + 3H_4SiO_4 \tag{13}$$

$$Al_2Si_2O_5(OH)_{4(s)} + 6H^+ \rightarrow 2Al^{3+} + 2H_4SiO_4 + H_2O \tag{14}$$

$$Ba(Al_2Si_2O_8)_{(s)} + 8H^+ \rightarrow Ba^{2+} + 2Al^{3+} + 2H_4SiO_4 \tag{15}$$

4.3 Geochemical processes in substrates

Sulfide oxidation is the dominant geochemical reaction occurring across all substrates, driving acidification. In addition, various acid neutralization mechanisms are at play, depending on the specific amendments applied. Siderite is only carbonate mineral existing in pure tailings without amendments. While the dissolution of siderite does not substantially neutralize the acidity overall, it is still regarded as a carbonate mineral. Those silicate minerals—muscovite, kaolinite, and celsian—also dissolve and neutralize the acidity, leading to the formation of secondary minerals such as quartz and jarosite.

In T+BR and T+BR+C, calcite dissolution is the first neutralization process, which is followed by siderite dissolution. Furthermore, goethite, böhmite, and gibbsite are Al&Fe (oxy)hydroxide minerals existing in bauxite residue, of which the dissolution would happen in the tailings mixing with bauxite residue. The weathering of silicate minerals involves the following processes, which include muscovite, celsian, and kaolinite in original tailings and sodalite in bauxite residue.

In T+L and T+L+C, the most notable change to tailings is the addition of lime, and calcite when exposed to air with CO_2 [45]. Therefore, the geochemical processes occurring in tailings amended with lime are the dissolution of calcium hydroxide ($Ca(OH)_2$) and calcite ($CaCO_3$), and those above-mentioned processes in tailings, such as dissolution of siderite and silicates weathering (muscovite, kaolinite, and celsian).

In T+C, T+BR+C, and T+L+C, the primary change is that the addition of large amounts of organic matter results in the decomposition of organic matters, leading to significant acidification. However, the organic matters decomposition is organic process falling out of scope of this research as the inorganic-related geochemical processes are the focus of this research, which, thus, will not be discussed in this chapter. Nevertheless, the discussion in Section 4.1 implies that the acidification by root exudates is the second organic process in the rhizosphere in pots with grasses.

Up to this point, the inorganic-related geochemical processes taking place in the rhizosphere of the various amended tailings have been thoroughly discussed. Even though the organic processes have not been explored in depth here, they still impact drainage pH, metal mobility within the substrates, and the reaction rates of

	T	T+BR	T+L	T+C	T+BR+C	T+L+C
Sulfides oxidation	Pyrite, galena, sphalerite	Pyrite, galena, sphalerite	Pyrite, galena, sphalerite	Pyrite, galena, sphalerite	Pyrite, galena, sphalerite	Pyrite, galena, sphalerite
Carbonates dissolution	Siderite	Calcite, siderite	Calcite, siderite	Siderite	Calcite, siderite	Calcite, siderite
(oxy) hydroxide dissolution	-	gibbsite, böhmite, goethite, hematite	calcium, hydroxide	-	gibbsite, böhmite, goethite, Hematite	calcium hydroxide
Silicates weathering	Muscovite, kaolinite, celsian	Muscovite, kaolinite, celsian, sodalite	Muscovite, kaolinite, celsian	Muscovite, kaolinite, celsian	Muscovite, kaolinite, celsian, sodalite	Muscovite, kaolinite, celsian

T = tailings; BR = bauxite residue; T = lime; C = compost.

Table 5.
Minerals responsible for different inorganic geochemical processes in the six substrates.

geochemical processes in several ways. All the minerals responsible for the various geochemical processes in the six substrates are summarized in **Table 5**.

5. Conclusion

In conclusion, inorganic geochemical reactions affecting the mobility of Pb and Zn were explored for the rhizosphere of phytostabilization in this chapter. These processes are grouped into two types: acidification processes and acid neutralization processes. The acidification includes primary sulfide oxidation, decomposition of organic matter, and the root exudates acidification. The acid neutralization processes occur sequentially through carbonate dissolution, Al and Fe (oxy)hydroxide dissolution, and silicate weathering.

While the alkaline materials were added to modify pH, both the end pH of the drainage and substrates in pots are quite acidic. This outcome was due to three acidification processes that increased metal mobility, as neutralization processes could not provide the lasting influence across the phytostabilization period. Upon the completion of a neutralization process or the depletion of a specific mineral, the pH drops quickly, and the subsequent neutralization reaction follows. Therefore, reapplying amendments regularly is essential for maintaining the long-term efficiency of phytostabilization, as it helps to sustain the neutralization processes.

It is suggested that exploring the combined effects of inorganic geochemical processes and organic processes on metal mobility would be the next step, as this research did not take microbial activity into account. Incorporating microbial effects could deepen our understanding of the interactions between tailings and roots during phytostabilization. Gaining insights into both of the inorganic and organic interactions would strengthen the knowledge foundation for future research on the fundamental reactive transport modeling of sulfidic mine tailings. Additionally, further research is advised at larger spatial scales, whether in greenhouses or field settings such as extending the duration of experiments and selecting species with longer lifespans. This approach would offer more detailed insights and patterns of geochemical processes, improving the implementation of phytostabilization during mine closure.

Acknowledgements

The authors express their appreciation to the Faro mine, UBC building operations, and Rio Tinto for providing tailings, compost, and bauxite residue, respectively. This research did not receive any specific funding from public, commercial, or not-for-profit agencies.

Author details

Lina Xie* and Dirk van Zyl
The Norman B. Keevil Institute of Mining Engineering, University of British Columbia, Vancouver, British Columbia, Canada

*Address all correspondence to: linaxie1992@gmail.com

References

[1] Lindsay MB, Moncur MC, Bain JG, Jambor JL, Ptacek CJ, Blowes DW. Geochemical and mineralogical aspects of sulfide mine tailings. Applied Geochemistry. 2015;**57**:157-177

[2] Park I, Tabelin CB, Jeon S, Li X, Seno K, Ito M, et al. A review of recent strategies for acid mine drainage prevention and mine tailings recycling. Chemosphere. 2019;**219**:588-606

[3] Trippe KM, Manning VA, Reardon CL, Klein AM, Weidman C, Ducey TF, et al. Phytostabilization of acidic mine tailings with biochar, biosolids, lime, and locally-sourced microbial inoculum: Do amendment mixtures influence plant growth, tailing chemistry, and microbial composition? Applied Soil Ecology. 2021;**165**:103962

[4] Xie L, van Zyl D. Distinguishing reclamation, revegetation and phytoremediation, and the importance of geochemical processes in the reclamation of sulfidic mine tailings: A review. Chemosphere. 2020;**252**:126446

[5] Zine H, Midhat L, Hakkou R, El Adnani M, Ouhammou A. Guidelines for a phytomanagement plan by the phytostabilization of mining wastes. Scientific African. 2020;**10**:e00654

[6] Farahat EA, Galal TM. Trace metal accumulation by *Ranunculus sceleratus*: implications for phytostabilization. Environmental Science and Pollution Research. 2018;**25**(5):4214-4222

[7] Garau M, Castaldi P, Diquattro S, Pinna MV, Senette C, Roggero PP, et al. Combining grass and legume species with compost for assisted phytostabilization of contaminated soils. Environmental Technology & Innovation. 2021;**22**:101387

[8] Shi Y, Zang Y, Yang H, Zhang X, Shi J, Zhang J, et al. Biochar enhanced phytostabilization of heavy metal contaminated mine tailings: A review. Frontiers in Environmental Science. 2022;**10**:1044921

[9] Xie L, van Zyl D. Identification of grass species candidates for phytostabilization and enhanced metal (loid) s immobilisation using cost-effective amendments on sulfidic mine tailings. International Journal of Mining, Reclamation and Environment. 2023:1-15

[10] Li Y, Sun Q, Zhan J, Yang Y, Wang D. Vegetation successfully prevents oxidization of sulfide minerals in mine tailings. Journal of Environmental Management. 2016;**177**:153-160

[11] Yang B, Luo W, Wang X, Yu S, Gan M, Wang J, et al. The use of biochar for controlling acid mine drainage through the inhibition of chalcopyrite biodissolution. Science of the Total Environment. 2020;**737**:139485

[12] Khan MY, Asghar HN, Jamshaid MU, Akhtar MJ, Zahir ZA. Effect of microbial inoculation on wheat growth and phytostabilization of chromium contaminated soil. Pakistan Journal of Botany. 2013;**45**(S1):27-34

[13] Lugtenberg BJ, Malfanova N, Kamilova F, Berg G. Plant growth promotion by microbes. Molecular Microbial Ecology of the Rhizosphere. 2013;**2**:561-573

[14] Li Y, Jia Z, Sun Q, Zhan J, Yang Y, Wang D. Ecological restoration alters microbial communities in mine tailings profiles. Scientific Reports. 2016;**6**:25193

[15] Radziemska M, Gusiatin ZM, Cydzik-Kwiatkowska A, Cerdà A,

Pecina V, Bęś A, et al. Insight into metal immobilization and microbial community structure in soil from a steel disposal dump phytostabilized with composted, pyrolyzed or gasified wastes. Chemosphere. 2021;**272**:129576

[16] Zhang MJ, Liu XY, Li YB, Wang GY, Wang ZN, Wen JK. Microbial community and metabolic pathway succession driven by changed nutrient inputs in tailings: Effects of different nutrients on tailing remediation. Scientific Reports. 2017;7:474

[17] Hagmann DF, Goodey NM, Mathieu C, Evans J, Aronson MF, Gallagher F, et al. Effect of metal contamination on microbial enzymatic activity in soil. Soil Biology and Biochemistry. 2015;**91**:291-297

[18] Honeker LK, Gullo CF, Neilson JW, Chorover J, Maier RM. Effect of re-acidification on buffalo grass rhizosphere and bulk microbial communities during phytostabilization of metalliferous mine tailings. Frontiers in Microbiology. 2019;**10**:1209

[19] Lee S-H, Ji W, Lee W-S, Koo N, Koh IH, Kim M-S, et al. Influence of amendments and aided phytostabilization on metal availability and mobility in Pb/Zn mine tailings. Journal of Environmental Management. 2014;**139**:15-21

[20] Wu S, Liu Y, Southam G, Robertson L, Chiu TH, Cross AT, et al. Geochemical and mineralogical constraints in iron ore tailings limit soil formation for direct phytostabilization. Science of the Total Environment. 2019; **651**:192-202

[21] Muthusamy L, Rajendran M, Ramamoorthy K, Narayanan M, Kandasamy S. Phytostabilization of metal mine tailings—A green remediation technology. In: Vineet K, Maulin S, Sushil S, editors. Phytoremediation Technology for the Removal of Heavy Metals and Other Contaminants from Soil and Water. Elsevier; 2022. pp. 243-253

[22] Dighton J, Krumins JA. Interactions in Soil: Promoting Plant Growth. Vol. 1. Springer; 2014

[23] Tadych M, Bergen MS, White JF Jr. Epichloë spp. associated with grasses: new insights on life cycles, dissemination and evolution. Mycologia. 2014;**106**(2): 181-201

[24] Thomas GW. Soil pH and soil acidity. In: Sparks DL, editor. Methods of Soil Analysis Part 3—Chemical Methods (methodsofsoilan3). Madison, WI: Soil Science Society of America; 1996. pp. 475-490

[25] Sobek AA. Field and Laboratory Methods Applicable to Overburdens and Minesoils: Industrial Environmental Research Laboratory. US Environmental Protection Agency: Office of Research and Development; 1978

[26] Lasaga AC. Fundamental approaches in describing mineral dissolution and precipitation rates. In: Chemical Weathering Rates of Silicate Minerals. De Gruyter; 2018. pp. 23-86

[27] Oelkers EH. General kinetic description of multioxide silicate mineral and glass dissolution. Geochimica et Cosmochimica Acta. 2001;**65**(21): 3703-3719

[28] Weber PA, Thomas JE, Skinner WM, Smart RSC. A methodology to determine the acid-neutralization capacity of rock samples. The Canadian Mineralogist. 2005;**43**(4):1183-1192

[29] Jambor JL, Dutrizac JE, Raudsepp M, Groat LA. Effect of Peroxide on Neutralization-Potential Values of

Siderite and Other Carbonate Minerals. Journal of Environmental Quality. 2003; **32**(6):2373-2378

[30] Paktunc AD. Characterization of mine wastes for prediction of acid mine drainage. In: Environmental Impacts of Mining Activities. Springer; 1999. pp. 19-40

[31] Baleeiro A, Fiol S, Otero-Fariña A, Antelo J. Surface chemistry of iron oxides formed by neutralization of acidic mine waters: Removal of trace metals. Applied Geochemistry. 2018;**89**:129-137

[32] Gunsinger M, Ptacek C, Blowes D, Jambor J, Moncur M. Mechanisms controlling acid neutralization and metal mobility within a Ni-rich tailings impoundment. Applied Geochemistry. 2006;**21**(8):1301-1321

[33] Brown A, Jurinak J. Mechanism of pyrite oxidation in aqueous mixtures. Journal of Environmental Quality. 1989; **18**(4):545-550

[34] Nordstrom DK, Alpers CN. Geochemistry of acid mine waters. The Environmental Geochemistry of Mineral Deposits. 1999;**6**(October):133-160

[35] Sumner ME, Noble AD. Soil acidification: the world story. In: Handbook of Soil Acidity. CRC Press; 2003. pp. 15-42

[36] Goulding K. Soil acidification and the importance of liming agricultural soils with particular reference to the United Kingdom. Soil Use and Management. 2016;**32**(3):390-399

[37] Haynes R. Soil acidification induced by leguminous crops. Grass and Forage Science. 1983;**38**(1):1-11

[38] Anderson DC, Jones SG. Clay barrier-leachate interaction. In: National Conference on Management of Uncontrolled Hazardous Waste Sites. Washington, D.C.: October 31-November 2, 1983. pp. 154-160

[39] Ulrich B. Natural and anthropogenic components of soil acidification. Zeitschrift für Pflanzenernährung und Bodenkunde. 1986;**149**(6):702-717

[40] Bolan NS, Hedley MJ. Role of carbon, nitrogen, and sulfur cycles in soil acidification. In: Handbook of Soil Acidity. CRC Press; 2003. pp. 43-70

[41] Gunsinger M, Ptacek C, Blowes D, Jambor J. Evaluation of long-term sulfide oxidation processes within pyrrhotite-rich tailings, Lynn Lake, Manitoba. Journal of Contaminant Hydrology. 2006;**83**(3-4):149-170

[42] Shingwenyana R, Shabalala AN, Mbhele R, Masindi V. Techno-economic analysis of the reclamation of drinking water and valuable minerals from acid mine drainage. Minerals. 2021;**11**(12):1352

[43] Johnson R, Blowes D, Robertson W, Jambor J. The hydrogeochemistry of the Nickel Rim mine tailings impoundment, Sudbury, Ontario. Journal of Contaminant Hydrology. 2000;**41**(1-2):49-80

[44] Jambor J, Blowes D. Theory and applications of mineralogy in environmental studies of sulfide-bearing mine wastes. Modern Approaches to Ore and Environmental Mineralogy. 1998;**27**: 367-401

[45] Cizer Ö, Van Balen K, Elsen J, Van Gemert D. Crystal morphology of the precipitated calcite crystals from accelerated carbonation of lime binders. In: Proceedings of ACEME08, 2nd International Conference on Accelerated Carbonation for Environmental and Materials Engineering. Rome; 1-3 October 2008. pp. 139-148

Chapter 4

Innovations in Herbicide Bioremediation: Green Solutions for Soil Contamination

Karolayne Silva Souza, Milena Roberta Freire da Silva, Kaline Catiely Campos Silva, Lívia Caroline Alexandre de Araújo, Matheus Victor Aragão Cavalcanti, Amanda Vieira de Barros, Jamile Maria Pereira Bastos Lira de Vasconcelos, Fabricio Motteran and Maria Betânia Melo de Oliveira

Abstract

This chapter discusses innovations in herbicide bioremediation as a sustainable solution for agricultural soil contamination. Modern agriculture, essential for global food security, faces significant environmental challenges due to the excessive use of herbicides, which can accumulate in the soil and negatively impact ecosystems and human health. Bioremediation emerges as a promising alternative, utilizing living organisms such as microorganisms and transgenic plants to degrade or transform these toxic compounds. This study explores how advances in genetic engineering and nanotechnology have enhanced the efficiency of these processes, making soil decontamination more effective and less harmful to the environment. Furthermore, it is acknowledged that, although this technology offers significant benefits, such as reducing environmental impacts and economic viability, challenges remain particularly in adapting to different climatic conditions and managing increasingly complex herbicides. With a focus on promoting safer and more sustainable agricultural practices, this study contributes to building agriculture that balances productivity with environmental preservation, emphasizing the importance of innovative solutions that can mitigate the damage caused by agrochemicals and protect ecosystems and public health.

Keywords: bioremediation, herbicides, degradation, sustainable agriculture, microorganisms

1. Introduction

In Brazil, between the 1940s and 1950s, the agricultural sector saw a significant increase in the use of synthetic fertilizers and pesticides, driven mainly by tax incentives, increased farmer incomes, and the provision of rural credit by the federal government. In this context, government policies and marketing strategies facilitated the widespread use of pesticides and fertilizers by both large-scale producers and family farmers [1].

Modern agriculture, while crucial for ensuring global food security, faces major environmental challenges, particularly due to the indiscriminate use of herbicides. These chemical compounds, essential for weed control, often accumulate in the soil, causing contamination and posing threats to both local ecosystems and human health. Bioremediation therefore emerges as a promising alternative to address these issues in a sustainable way [2].

By utilizing microorganisms and plants to degrade herbicides, bioremediation offers a more ecological approach compared to conventional methods for treating contaminated soils. With advancements in technologies such as genetic engineering and nanotechnology, the efficiency of this process has been significantly enhanced. However, challenges remain, including the need for adaptation to different environmental conditions and the complexity of modern herbicides [3, 4].

This research aims to explore the latest innovations in bioremediation, with an emphasis on techniques that promote agricultural sustainability and reduce the risks of soil contamination.

2. General characteristics of agriculture

Agriculture is considered a comprehensive economic activity that involves both the cultivation of plants and the raising of animals, primarily for food production and essential goods that sustain the global human population [5]. As such, the main characteristics that define agriculture are its economic, technological, biological, and environmental aspects, which vary according to the region, climate, and the adoption of different cultural practices [6].

Historically, the transformation of agriculture occurred primarily through human evolution over time, as food needs increased. Consequently, it became essential to develop strategies for land use to establish a system of food cultivation [7].

In this context, agriculture can be classified into two main types: commercial agriculture, practiced by small rural properties aimed at meeting the family's food needs, with little or no production for commercialization, and subsistence agriculture, which is based on large-scale food production using advanced methods and techniques, with the primary goal of generating profit [8]. Additionally, agricultural production systems can be classified as monoculture or polyculture. Monoculture refers to the cultivation of a single species over a large area, while polyculture involves the practice of cultivating multiple species in the same area [9].

Agriculture, despite its specific characteristics, is still highly dependent on the global economy, state action, and current public policies. One of the most significant changes in global agriculture was the Green Revolution, which provided genetic improvement of seeds, the use of fertilizers, machinery, and pesticides, resulting in a more capitalized agriculture, the development of Agro-industrial Complexes (CAIs), and an increase in food productivity [7].

The international food market is considered a complex and dynamic environment, as it depends on global demands and trends. In this context, due to the significant population growth and urbanization, the market has been constantly influenced, particularly by the increasing demand for food [10].

Countries such as the United States, China, Germany, Japan, and the United Kingdom are the main representatives of food importers, relying on external supplies to meet their demands. On the other hand, the main food-exporting countries include Brazil, the United States, Canada, Australia, and Argentina, which have been supplying agricultural products to the global market and driving demand for higher value-added foods [11].

For this reason, greater efficiency and modernization in agriculture require the implementation of agricultural technologies, which primarily involve mechanization through the use of agricultural machinery, biotechnology that includes methods for pest-resistant crops, and pesticides aimed at protecting crops from pests, diseases, and weeds [12, 13].

One of the main challenges faced by agriculture is food security, which seeks to ensure global food demand, along with sustainability, which is focused on environmental preservation and biodiversity conservation to ensure long-term economic viability [14]. As a result, agriculture has significantly impacted the environment, mainly due to the use of natural resources such as water, energy, and soil [15, 16].

These resources are essential for agricultural activities, and their sustainable management is crucial to preventing environmental degradation and mitigating global climate impacts [17]. Therefore, it is of utmost importance that countries in the international food market work toward meeting global demands and trends to enhance their food production capacity and manage the fluctuations in agricultural product prices, ensuring food quality and safety [10].

Thus, the adoption of sustainable agricultural practices can contribute to biodiversity conservation, improve soil and water quality, and reduce greenhouse gas emissions. Examples include the implementation of integrated management techniques, the use of organic fertilizers, and the preservation of native vegetation areas [18, 19].

2.1 Brazil's position in relation to the world

Brazil is considered one of the largest producers in the global agricultural landscape, primarily distinguished by its vast territorial extension, climatic diversity, and fertile soils, which enable both family farming and large-scale agriculture, along with a significant variety of crops [20, 21]. As such, agribusiness has played a significant role in the country's economic growth, being responsible for the majority of both national and international food exports and supplies [11].

Agribusiness encompasses agricultural, livestock, fishing, and forestry activities, including everything from primary food production to distribution, agroindustry, and export of these products. Brazil stands out as one of the largest producers and exporters of agricultural goods. Among the various products produced and exported, soybeans, coffee, sugar, beef, and poultry are particularly important, being fundamental to global food supply [10, 22].

However, this expansion has posed significant challenges for the country, such as the deforestation of Brazilian biomes like the Amazon, a matter of global concern due to the relationship between extensive livestock farming and agriculture [23, 24]. In addition, the use of pesticides also presents challenges, as high quantities are used to ensure high productivity and the protection of commodities against pests and diseases [25, 26].

3. Alternatives used in Pest and disease control

In a balanced environment, plants can coexist with sufficient quantity and quality of soil nutrients, enabling them to produce complex substances such as proteins, fats, and minerals, without pests and diseases causing damage to their production. An environment is considered balanced when it has favorable conditions of moisture, light, and the presence of biological controllers and/or predators of diseases and pests, preventing the population of pests from increasing [27].

The rise of pathogens and pests in commodities has led to significant losses in agricultural crop production worldwide. Furthermore, the demand for food has intensified throughout the twentieth century, aligned with the growth of the global population, which is estimated to reach 8.5 billion by 2030, 9.7 billion by 2050, and 10.9 billion by 2100 [28].

Since the Green Revolution, agricultural production has increasingly relied on substantial use of inorganic fertilizers, synthetic pesticides, and genetically modified organisms to meet the growing food demand. Consequently, these alternatives have been adopted with the primary goal of quickly addressing pest management in commodities [29].

3.1 Natural and sustainable alternatives

Currently, natural alternatives have become a global trend to ensure healthy and adequate planting while also reducing the risks of environmental pollution and even the intoxication of operators and consumers. As a result, the planning of ecologically-based agricultural management has become increasingly necessary to ensure that food production is free from pesticides [27].

Conventional agriculture uses synthetic fertilizers, pesticides, and growth regulators to achieve higher crop yields and food production. In contrast, organic agriculture relies on natural and sustainable resources, such as biological pest control agents, the application of biofertilizers, crop rotation, and biopesticides [30, 31].

Biological control agents in agriculture are related to natural or genetically modified organisms used to reduce the impact of undesirable organisms. They utilize beneficial crops, insects, and microorganisms such as bacteria and fungi to induce resistance against unwanted pathogens in commodities and compete for space and nutrients with these organisms, acting through an antibiosis mechanism. These agents are considered alternatives for disease management in agriculture, particularly when compared to techniques involving chemical fertilizers and pesticides for pest control [32].

Biofertilizers aim to restore soil health and fertility by adding organic matter, as this practice improves the physical, chemical, and biological characteristics of the soil. Fertilizers may include crop residues from cultivated soils, animal manure, and compost [33, 34].

Crop rotation, on the other hand, aims to alternate different crop groups, such as cereals, legumes, and vegetables, optimizing soil productivity and improving its quality while reducing the risks of pests, diseases, or nutrient depletion [35]. An example of crop rotation is legumes, which can naturally fix nitrogen in the soil and help subsequent crops in pest and disease control, reducing the need for chemical pesticides [36].

In this context, biopesticides, which are also a sustainable agricultural practice and an alternative to synthetic pesticides, primarily utilize plant-based Essential Oils (EOs) and homemade pesticides [30]. EOs are derived from the nonwoody parts of plants, particularly the foliage, and are extracted through steam capture or hydrodistillation [37]. Homemade pesticides, which can be produced at home, are based on the preparation of natural products to repel pests and diseases [38].

3.2 Synthetic pesticides

In conventional management systems, one of the main forms of pest control is through chemical compounds such as pesticides, which are considered major contributors to environmental problems, including water, soil, and food contamination, as well as intoxication and other issues [27]. Pesticides are chemical compounds designed to control, destroy, repel, attract, or prevent any biological organism that may cause pests in crops. These compounds differ from one another based on their physical and chemical properties [39].

During the 1930s, there was an increase in the use of synthetic chemical pesticides, which included compounds such as arsenic and sulfur in formulations for crop protection. Following this, at the beginning of World War II, various pesticides were produced, such as Dichlorodiphenyltrichloroethane (DDT), aldrin, and dieldrin, used as insecticides, and 2-methyl-4-chlorophenoxyacetic acid (MCPA) and 2,4-dichlorophenoxyacetic acid (2,4-D) as herbicides [40].

In this context, pesticides have become an essential tool over the years for plant protection and crop improvement in the development of agricultural practices. This improvement is related to the management of pests, diseases, and weeds, with key categories including herbicides, insecticides, fungicides, rodenticides, nematicides, and other pesticides [41].

According to the United Nations Food and Agriculture Organization (FAO), from 1990 to 2021, large quantities of pesticides were recorded globally, especially in the Americas and Asia. The data shows that the global average amount of pesticides increased from 1.22 kg per hectare in 1990 to 2.26 kg per hectare in 2021 [42]. **Figure 1B** shows the percentage of pesticide use by country and continent.

Pesticides can also be classified based on criteria such as toxicity, target organism, and chemical composition [41, 43, 44]. According to the World Health Organization (WHO), pesticide classification based on toxicity corresponds to the Median Lethal Dose (LD50) value, expressed in mg/kg of body weight. In this way, pesticides are classified into five categories: Class Ia (Extremely Hazardous), Class Ib (Highly Hazardous), Class II (Moderately Hazardous), Class III (Slightly Hazardous), and Class U (Unlikely to Present Acute Hazard) [45], as described in **Table 1**.

The classification by target organism corresponds to fungicides, insecticides, herbicides, and rodenticides. Fungicides are associated with the inhibition or elimination of fungal growth that causes plant diseases, insecticides are used to repel or kill insects that affect agricultural crops, herbicides are applied to control and eradicate weeds, and rodenticides are used to control rodents in plantations or food storage areas [46, 47].

As for the chemical composition of pesticides, this is one of the most well-known classifications, organized according to their chemical properties. Fungicides are divided into groups such as aliphatic nitrogen, amides, aromatics, dicarboximides, and dinitrophenolics. Herbicides, in turn, are classified as anilides, phenoxyacetic acids, quaternary ammonium compounds, chlorotriazines, and sulfonylureas. Insecticides are categorized as carbamates, organochlorines, organophosphates, pyrethroids, neonicotinoids, spinosyns, benzoylureas, and antibiotics. Rodenticides are classified as inorganic, including zinc and aluminum phosphide, and organic, such as bromadiolone and coumatetralyl [41, 48].

In this context, pesticides are chemical compounds regulated by laws, regulations, and guidelines worldwide, subject to approval, registration, and entry into the

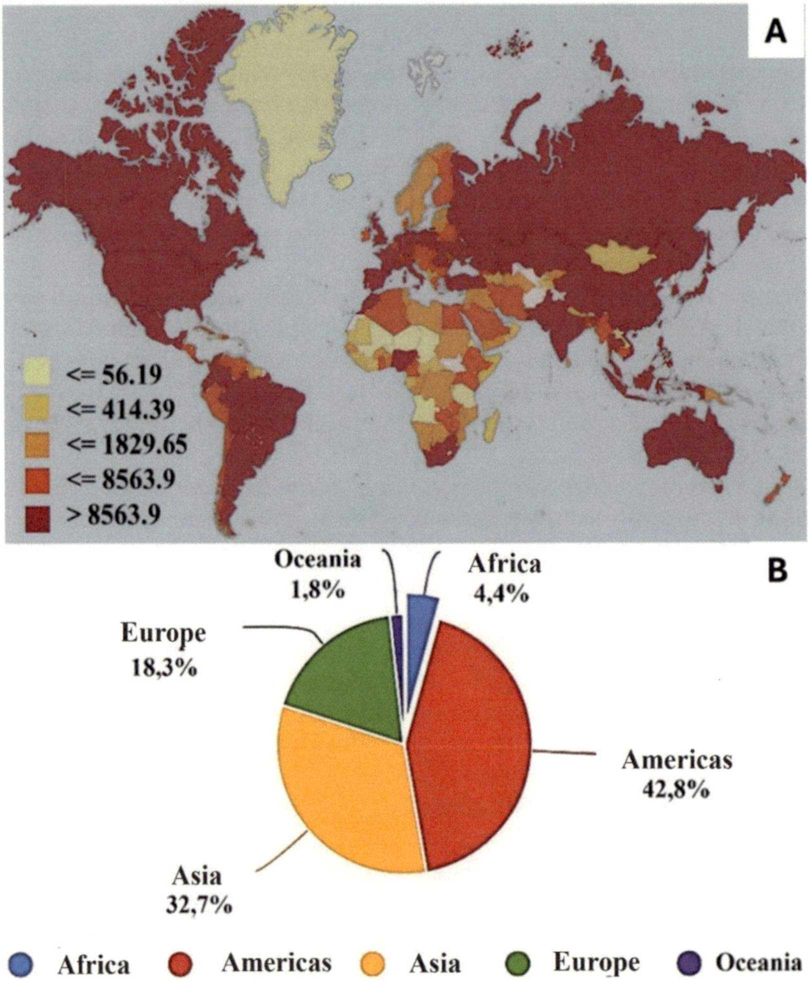

Figure 1.
Pesticide use worldwide (A) and by continent (B). Source: Adapted, FAO [42].

Class	Description	Oral LD50 (mg/kg)	LD50 dermal (mg/kg)
Ia	Extremely hazardous	<5	<50
Ib	Highly hazardous	5–50	50–200
II	Moderately hazardous	50–2000	200–2000
III	Slightly hazardous	>2000	>2000
U	Unlikely to present acute hazard	≥5000	≥5000

Table 1.
Pesticide classification based on acute toxicity according to LD50 values. Source: Adapted, WHO [45].

industry. Even with established usage requirements, pesticides have characteristics that allow them to persist in the environment, as well as properties of bioaccumulation and toxicity for both the environment and humans, which makes their use hazardous [49].

4. General characteristics of herbicides

Herbicides are Class III chemical compounds widely used for controlling unwanted plants, especially weeds, which compete with agricultural crops for nutrients, light, and water. They play an essential role in modern agriculture by ensuring productivity and food security on a global scale. However, their use must be carefully managed to avoid negative impacts on the environment and human health [50, 51].

Herbicides work through mechanisms that interfere with vital plant processes such as photosynthesis, cell division, and amino acid synthesis. These compounds can be selective, affecting only specific types of plants, or nonselective, killing a wide range of plant species. Therefore, the efficacy of a herbicide depends on its ability to reach the target plant, be absorbed, translocated to the site of action, and ultimately cause sufficient damage to inhibit growth or induce death [52, 53].

Herbicides are also classified based on the timing of their application. Preemergent herbicides are applied to the soil before the germination of unwanted plants, creating a chemical barrier that prevents their growth. Post-emergent herbicides, on the other hand, are applied directly to plants, eliminating them after germination [54].

Another fundamental aspect of herbicides is their interaction with the environment. For instance, water solubility affects the dispersion of compounds in soil and water, influencing both their effectiveness and potential for environmental contamination [55]. In this context, highly soluble herbicides tend to move more easily, potentially reaching groundwater and bodies of water, while those with low solubility remain in the soil longer, increasing the risk of bioaccumulation. This also leads to accumulation in living organisms, which can have long-term ecological and toxicological consequences [56, 57].

The indiscriminate and repeated use of herbicides can lead to the development of resistance in weeds, making their control increasingly difficult. This requires the adoption of integrated pest management strategies, which include crop rotation, alternating different herbicides, and the development of new compounds that are more selective and less harmful to the environment [2, 58].

5. Bioremediation: Concept and advantages

Bioremediation is a technology that uses living organisms, such as bacteria, fungi, and plants, to degrade, transform, or remove contaminants from the environment. These organisms operate through natural biochemical processes that convert pollutants into less toxic or harmless substances, helping to restore degraded ecosystems [59]. Additionally, bioremediation can be applied to a wide range of pollutants, including organic compounds such as hydrocarbons and pesticides, as well as heavy metals [60, 61].

One of the greatest advantages of bioremediation is its sustainable approach. Compared to conventional remediation methods, such as incineration or chemical processes, bioremediation is less invasive and often more cost-effective [62, 63]. Another benefit is the ability to treat large areas efficiently, especially when *in situ* methods (at the contamination site) are used, avoiding the need to remove contaminated soil or water for external treatment [63].

Bioremediation can be classified into two main types: *in situ* and *ex situ*. The first occurs directly at the contaminated site, without the need to transport the material, making the process cheaper and less disruptive. *Ex situ* bioremediation, on the other hand, involves the removal of contaminated soil or water for treatment at another location, being more suitable for areas with severe or difficult-to-access contamination [64].

Furthermore, bioremediation can be enhanced through techniques such as biostimulation, which involves adding nutrients or agents that promote the growth and activity of degrading microorganisms, or bioaugmentation, which introduces specialized microorganisms to the contaminated site to accelerate the degradation process [65].

Despite its benefits, bioremediation has some limitations. Not all pollutants can be easily degraded by living organisms, and environmental conditions, such as temperature, pH, and oxygen availability, can affect the effectiveness of the process. However, with advances in genetic engineering and biotechnology, new microorganisms are being developed to expand the scope of bioremediation, making it a promising solution for the recovery of contaminated environments [65, 66].

5.1 Recent innovations in herbicide bioremediation

In recent years, the field of herbicide bioremediation has advanced considerably, driven by the development of new technologies and the growing demand for more sustainable agricultural practices [67]. These innovations are transforming the way herbicides are managed in the environment, offering more effective and ecologically responsible methods for the decontamination of soil and water. By using microorganisms and plants to degrade toxic substances, bioremediation has proven to be a promising solution to mitigate the impact of herbicides on natural ecosystems [65, 68].

One of the most significant innovations is the use of genetic engineering to modify microorganisms, such as bacteria and fungi, to optimize their ability to degrade herbicides [69]. These genetically modified microorganisms can break down complex herbicide molecules into less harmful compounds, accelerating the decontamination process. An important example is the development of bacteria that can rapidly metabolize herbicides like glyphosate and atrazine, reducing their toxic impacts and preventing the accumulation of residues in the soil [70–72].

In addition to microorganisms, transgenic plants have also played an important role in new bioremediation techniques. Phytoremediation, which uses plants to absorb and degrade pollutants, has been improved by the introduction of genes that enhance these plants' ability to process and neutralize herbicides. These plants are particularly effective in agricultural environments where herbicides are widely used and can accumulate in the soil, threatening local biodiversity [73, 74].

The application of phytoremediation techniques, in combination with microorganisms living in plant roots (rhizobacteria), has also shown promising results in the degradation of herbicides. This symbiotic interaction enhances the plants' ability to absorb and transform pollutants, speeding up the recovery of contaminated soils, particularly in agricultural areas [75].

A complementary approach gaining attention is nanoparticle-assisted bioremediation, which uses nanoparticles to transport microorganisms or enzymes directly to contaminated sites, increasing degradation efficiency. This technique allows microorganisms to reach difficult-to-access areas and interact more rapidly with pollutants, optimizing the bioremediation process and reducing the time needed for environmental recovery [76, 77].

The use of biosurfactants also represents a crucial advance in herbicide bioremediation. These biological compounds, produced by microorganisms, increase the solubility of herbicides in the soil, facilitating their biodegradation. By reducing the surface tension between the pollutant and the environment, biosurfactants improve the interaction between microorganisms and herbicides, accelerating the breakdown of toxic compounds [78, 79].

Another important development is the improvement of biostimulation techniques, which involve providing nutrients and optimal conditions to promote the growth and activity of degrading microorganisms at the contaminated site. By creating a favorable environment, biostimulation can increase the rate of herbicide degradation, making the process more efficient and applicable to large agricultural areas [80].

Bioaugmentation is another innovative technique that involves introducing specific microorganisms into contaminated areas. Microorganisms that exhibit high efficiency in herbicide degradation are added to the soil or water, complementing the action of the microorganisms already present in the environment. This is particularly useful in areas where the natural microbial populations are not capable of fully degrading the pollutants [80].

6. Future perspectives and sustainability

Future perspectives for herbicide bioremediation are increasingly focused on innovative strategies that not only aim to enhance the process's effectiveness but also ensure its long-term sustainability. With the growing environmental impact caused by the extensive use of herbicides in agriculture, the demand for solutions that can restore ecosystems safely and efficiently has intensified [81].

One of the most promising trends is the use of advanced biotechnology to genetically modify microorganisms, making them more efficient at degrading complex herbicides. Genetic modification allows microorganisms to develop new metabolic pathways or increase the expression of enzymes responsible for degradation, making the bioremediation process more targeted and effective [82, 83].

The integration of molecular monitoring technologies is transforming the way bioremediation is applied and evaluated, as next-generation genetic sequencing allows real-time monitoring of herbicide degradation progress. This enables scientists to adjust remediation conditions as needed, making the process more dynamic and effective [84, 85].

The future of bioremediation must also consider adaptation to climate change, as the effectiveness of microorganisms is highly dependent on environmental conditions such as temperature and humidity. Research aimed at developing strains more resilient to these climatic variations is crucial. This will ensure that bioremediation techniques remain effective even in scenarios of extreme environmental changes [65].

7. Conclusion

Herbicide bioremediation represents an innovative and sustainable approach to mitigating the environmental impacts caused by the intensive use of agrochemicals. Recent innovations, such as the genetic modification of microorganisms and the use of transgenic plants, have enabled the development of more effective solutions for herbicide degradation, reducing their persistence in soil and water.

The introduction of technologies like nanotechnology and biosurfactants has also improved the efficiency of the process by increasing the availability of herbicides to biodegrading agents and facilitating the remediation of contaminated areas. These strategies, in addition to being economically viable, offer a less harmful alternative to the environment compared to conventional methods of soil and water treatment.

Bioremediation stands out for using natural biological processes, minimizing the generation of toxic by-products and promoting a cleaner environmental recovery. However, for these solutions to become widely adopted, challenges related to environmental variations and climate change must be addressed, as they can affect the effectiveness of microorganisms in the field.

Although herbicides are essential for modern agriculture, their use requires a balanced approach. It is crucial to follow best management practices to minimize negative impacts and ensure that these substances fulfill their role without compromising the health of ecosystems and human populations. Continuous innovation in the development of more sustainable herbicides and the implementation of strict regulations are important steps in this direction.

Acknowledgements

We would like to thank the Coordination for the Improvement of Higher Education Personnel (CAPES) for the financial and institutional support provided, which was essential for carrying out this research.

Conflict of interest

The authors declare no conflict of interest.

Author details

Karolayne Silva Souza[1*], Milena Roberta Freire da Silva[1], Kaline Catiely Campos Silva[2], Lívia Caroline Alexandre de Araújo[1], Matheus Victor Aragão Cavalcanti[3], Amanda Vieira de Barros[1], Jamile Maria Pereira Bastos Lira de Vasconcelos[1], Fabricio Motteran[1] and Maria Betânia Melo de Oliveira[1]

1 Federal University of Pernambuco, Recife, Pernambuco, Brazil

2 University of the State of Bahia, Paulo Afonso, Bahia, Brazil

3 Frassinetti College of Recife, Recife, Pernambuco, Brazil

*Address all correspondence to: karolayne.silvasouza@ufpe.br

References

[1] Lima IB, Boëchat IG, Gücker B. Glyphosate in Brazil: Use, aquatic contamination, environmental effects and hazards to human health. Caderno de Geografia. 2021;**31**(e1):90-115

[2] Ofosu R, Agyemang ED, Márton A, Pásztor G, Taller J, Kazinczi G. Herbicide resistance: Managing weeds in a changing world. Agronomy. 2023;**13**(6):1595

[3] Singh Y, Saxena MK. Insights into the recent advances in nano-bioremediation of pesticides from the contaminated soil. Frontiers in Microbiology. 2022;**13**:982611

[4] Ahmar S, Mahmood T, Fiaz S, Mora-Poblete F, Shafique MS, Chattha MS, et al. Advantage of nanotechnology-based genome editing system and its application in crop improvement. Frontiers in Plant Science. 2021;**12**:663849

[5] Steensland A, Zeigler M. Productivity in Agriculture for a Sustainable Future. 2021

[6] Çakmakçı R, Salık MA, Çakmakçı S. Assessment and principles of environmentally sustainable food and agriculture systems. Agriculture. 2023;**13**(5):1073

[7] Castanho RB, Teixeira MES. The evolution of agriculture in the world: From its genesis to the present day. Brazilian Geographical Journal: Geosciences and Humanities Research Medium, Ituiutaba. 2017;**8**(1):136-146

[8] Macarringue AMJS, Castillo RA. Public Agricultural Policies and Their Implications On Family Farming In Mozambique. 2021

[9] Arsyad M, Sabang Y, Agus N, Bulkis S, Kawamura Y. Intercropping farming system and farmers income. AGRIVITA Journal of Agricultural Science. 2020;**42**(2):360-366

[10] Vieira LS. Agribusiness: The Development of the Sector with Little Impact and Great Recovery. 2023

[11] Oliveira EC, Carraro NC. Analysis of the behavior and participation of agribusiness in the composition of the Brazilian gross domestic product (GDP): A time series study from 1996 to 2017. Brazilian Journal of Development. 2019;**5**(11):24042-24064

[12] Ghosh PK, Das A, Saxena R, Banerjee K, Kar G, Vijay D, et al. Trajectory of 75 Years of Indian Agriculture after Independence. Springer; 2023

[13] Rehman A, Jingdong L, Khatoon R, Hussain I, Iqbal MS. Modern agricultural technology adoption its importance, role and usage for the improvement of agriculture. Life Science Journal. 2016;**14**(2):70-74

[14] Amorim JODL. Community Supported Agriculture (CSA) in São Paulo and Solidarity Agriculture (Solawi) in Germany: Building Social, Economic and Environmental Indicators. 2018

[15] Xu Z, Chen X, Liu J, Zhang Y, Chau S, Bhattarai N, et al. Impacts of irrigated agriculture on food-energy-water-CO2 nexus across metacoupled systems. Nature Communications;**11**(1):5837

[16] Wassie SB. Natural resource degradation tendencies in Ethiopia: A review. Environmental Systems Research. 2020;**9**(1):1-29

[17] Srivastav AL, Dhyani R, Ranjan M, Madhav S, Sillanpää M. Climate-resilient strategies for sustainable management of water resources and agriculture. Environmental Science and Pollution Research. 2021;**28**(31):41576-41595

[18] Ferraz RPD, Prado RB, Parron LM, Campanha MM. Framework for Ecosystem Services. Brasília, DF: Embrapa; 2019

[19] Cordeiro LAM, Vilela L, Marchão RL, Kluthcouski J, Martha Júnior GB. Crop-Livestock Integration and Crop-Livestock-Forest Integration: Strategies for Sustainable Intensification of Land Use. 2015

[20] Marin FR, Pilau FG, Spolador HF, Otto R, Pedreira CG. Sustainable intensification of Brazilian agriculture: Scenarios for 2050. Journal of Agricultural Policy. 2016;**25**(3):108-124

[21] Santos AF. Trends and challenges. A study of the perceptions of rural producers and agribusiness representatives about the future of Brazilian agribusiness. Competitiveness and Sustainability Journal. 2023;**10**(1):81-104

[22] Sabão DE. Socioeconomic Panorama of Agribusiness in Western Bahia. Barreiras: AIBA; 2015

[23] Campagnolla C, Macêdo MMC. Green revolution: Past and current challenges. Science & Technology Notebooks. 2022;**39**(1):26952

[24] Santos GRD, Silva RPD, Santana ASD. Agriculture in the Amazon: Deforestation, Scale and Challenges to Sustainable Production. 2022

[25] Lignani LDB, Brandão JLG. The dictatorship of pesticides: The National Agricultural Defensives Program and changes in pesticide production and consumption in Brazil, 1975-1985. History, Sciences, Health-Manguinhos. 2022;**29**(2):337-359

[26] Lucas, de Souza J, Alves dos Santos J, Cristina Marjotta-Maistro M, Sanjuan Montebello AE. Main active ingredients of agrochemicals: Marketing and use in Brazil. GeSec: Journal of Management and Secretariat. 2023;**14**(6)

[27] Junior JSZ, Lazzarini AL, de Oliveira AA, Rodrigues LA, de Moraes Souza II, Andrikopoulos FB, et al. Agroecological pest management: Alternatives for sustainable agriculture. Intelletto Scientific Journal. 2018;**3**(3)

[28] Savary S, Willocquet L, Pethybridge SJ, Esker P, McRoberts N, Nelson A. The global burden of pathogens and pests on major food crops. Nature Ecology & Evolution. 2019;**3**(3):430-439

[29] Ngegba PM, Cui G, Khalid MZ, Zhong G. Use of botanical pesticides in agriculture as an alternative to synthetic pesticides. Agriculture. 2022;**12**(5):600

[30] Durán-Lara EF, Valderrama A, Marican A. Natural organic compounds for application in organic farming. Agriculture. 2020;**10**(2):41

[31] Sumberg J, Giller KE. What is 'conventional' agriculture? Global Food Security. 2022;**32**:100617

[32] Singh S, Kumar V, Dhanjal DS, Singh J. Biological control agents: Diversity, ecological significances, and biotechnological applications. In: Natural Bioactive Products in Sustainable Agriculture. 2020. pp. 31-44

[33] Tejada M, Rodríguez-Morgado B, Gómez I, Franco-Andreu L, Benítez C, Parrado J. Use of biofertilizers obtained

from sewage sludges on maize yield. European Journal of Agronomy. 2016;**78**:13-19

[34] Beltrán-Pineda ME, Bernal-Figueroa AA. Biofertilizantes: Alternativa biotecnológica para los agroecosistemas. Revista Mutis. 2022;**12**(1)

[35] Barbosa JRV, Castilho JV, Rossato OB. Crop rotation. In: Proceedings of the Science, Technology, Art and Culture Fair of the Federal Institute of Santa Catarina, Concórdia Campus. Vol. 6(1). 2023. pp. 58-58

[36] Nharreluga ÉA. Fertilizer value of annual legumes grown as green manure [master's thesis]. Portugal: Instituto Politecnico de Braganca; 2023

[37] Azeem M, Zaman T, Tahir M, Haris A, Iqbal Z, Binyameen M, et al. Chemical composition and repellent activity of native plants essential oils against dengue mosquito, *Aedes aegypti*. Industrial Crops and Products. 2019;**140**:111609

[38] Rodrigues S, Diogo BS, Antunes SC, Lourenço I. Biopesticides: Challenges for research. Elementary Science Journal. 2023;**11**(2)

[39] Abubakar Y, Tijjani H, Egbuna C, Adetunji CO, Kala S, Kryeziu TL, et al. Pesticides, history, and classification. In: Natural Remedies for Pest, Disease and Weed Control. Academic Press; 2020. pp. 29-42

[40] Raffa CM, Chiampo F. Bioremediation of agricultural soils polluted with pesticides: A review. Bioengineering. 2021;**8**(7):92

[41] Nayak P, Solanki H. Pesticides and Indian agriculture—A review. International Journal of Research Granthaalayah. 2021;**9**(5):250-263

[42] FAO (Food and Agriculture Organization). FAOSTAT. 2024. Available from: https://www.fao.org/faostat/en/?#data/RP/visualize [Accessed: September 2, 2024]

[43] Tudi M, Daniel Ruan H, Wang L, Lyu J, Sadler R, Connell D, et al. Agriculture development, pesticide application and its impact on the environment. International Journal of Environmental Research and Public Health. 2021;**18**(3):1112

[44] Hassaan MA, El Nemr A. Pesticides pollution: Classifications, human health impact, extraction and treatment techniques. The Egyptian Journal of Aquatic Research. 2020;**46**(3):207-220

[45] World Health Organization (WHO). The WHO Recommended Classification of Pesticides by Hazard and Guidelines to Classification 2019. World Health Organization; 2020

[46] Nascimento JHA, Xavier GR, Pinto GDOA, Freitas IDA, Teixeira MN. Emergency treatment in cases of aldicarb poisoning—Literature review. Multidisciplinary Journal of Health. 2021;**2**(3):08

[47] Florido ALL. Pesticides: A Review of their History, Classification, Uses and Environmental Impacts with a Focus on Agroecology and Open Innovation as Sustainable Alternatives. 2021

[48] Constantin PP, Gonçalves LYA. Biodegradation and Bioremediation of Pesticides: Bacterial Response Systems to Pesticide Exposure. 2022

[49] Lykogianni M, Bempelou E, Karamaouna F, Aliferis KA. Do pesticides promote or hinder sustainability in agriculture? The challenge of sustainable use of pesticides in modern agriculture. Science of the Total Environment. 2021;**795**:148625

[50] Cobb AH. Herbicides and Plant Physiology. John Wiley & Sons; 2022

[51] Pakdaman Sardrood B, Mohammadi Goltapeh E. Weeds, herbicides and plant disease management. In: Sustainable Agriculture Reviews 31: Biocontrol. 2018. pp. 41-178

[52] Hess FD. Herbicide effects on plant structure, physiology, and biochemistry. In: Pesticide Interactions in Crop Production. CRC Press; 2018. pp. 13-34

[53] Fedtke C. Biochemistry and Physiology of Herbicide Action. Springer Science & Business Media; 2012

[54] Wolter DA. Pre-Emergent and Post-Emergent Herbicide Performance on *E. tristachya* in California Orchards. Davis: University of California; 2021

[55] Mehdizadeh M, Mushtaq W, Siddiqui SA, Ayadi S, Kaur P, Yeboah S, et al. Herbicide residues in agroecosystems: Fate, detection, and effect on non-target plants. Reviews in Agricultural Science. 2021;**9**:157-167

[56] Yang C, Lim W, Song G. Reproductive toxicity due to herbicide exposure in freshwater organisms. Comparative Biochemistry and Physiology Part C: Toxicology & Pharmacology. 2021;**248**:109103

[57] Yadav H, Kumar R, Sankhla MS. Residues of pesticides and heavy metals in crops resulting in toxic effects on living organism. Journal of Seybold Report. 2020;**1533**:9211

[58] Souza Z. Weed management and control in apple orchards. Technical Bulletin. 2021;**202**

[59] Kumar A, Bisht BS, Joshi VD, Dhewa T. Review on bioremediation of polluted environment: A management tool. International Journal of Environmental Sciences. 2011;**1**(6):1079-1093

[60] Chen M, Xu P, Zeng G, Yang C, Huang D, Zhang J. Bioremediation of soils contaminated with polycyclic aromatic hydrocarbons, petroleum, pesticides, chlorophenols and heavy metals by composting: Applications, microbes and future research needs. Biotechnology Advances. 2015;**33**(6):745-755

[61] Zhang H, Yuan X, Xiong T, Wang H, Jiang L. Bioremediation of co-contaminated soil with heavy metals and pesticides: Influence factors, mechanisms and evaluation methods. Chemical Engineering Journal. 2020;**398**:125657

[62] Singh S, Gupta VK. Biodegradation and bioremediation of pollutants: Perspectives strategies and applications. International Journal of Pharmacology and Biological Sciences. 2016;**10**(1):53

[63] Bala S, Garg D, Thirumalesh BV, Sharma M, Sridhar K, Inbaraj BS, et al. Recent strategies for bioremediation of emerging pollutants: A review for a green and sustainable environment. Toxics. 2022;**10**(8):484

[64] Hussain A, Rehman F, Rafeeq H, Waqas M, Asghar A, Afsheen N, et al. In-situ, ex-situ, and nano-remediation strategies to treat polluted soil, water, and air—A review. Chemosphere. 2022;**289**:133252

[65] Sharma I. Bioremediation techniques for polluted environment: Concept, advantages, limitations, and prospects. In: Trace Metals in the Environment-New Approaches and Recent Advances. London, UK: IntechOpen; 2020

[66] Mishra M, Singh SK, Kumar A. Environmental factors affecting the

bioremediation potential of microbes. In: Microbe Mediated Remediation of Environmental Contaminants. Woodhead Publishing; 2021. pp. 47-58

[67] Gonçalves CR, da Silva Delabona P. Strategies for bioremediation of pesticides: Challenges and perspectives of the Brazilian scenario for global application—A review. Environmental Advances. 2022;**8**:100220

[68] Mehrotra T, Dev S, Banerjee A, Chatterjee A, Singh R, Aggarwal S. Use of immobilized bacteria for environmental bioremediation: A review. Journal of Environmental Chemical Engineering. 2021;**9**(5):105920

[69] Rafeeq H, Afsheen N, Rafique S, Arshad A, Intisar M, Hussain A, et al. Genetically engineered microorganisms for environmental remediation. Chemosphere. 2023;**310**:136751

[70] King RB, Sheldon JK, Long GM. Practical Environmental Bioremediation: The Field Guide. CRC Press; 2023

[71] Kaur R, Singh D, Kumari A, Sharma G, Rajput S, Arora S, et al. Pesticide residues degradation strategies in soil and water: A review. International Journal of Environmental Science and Technology. 2021:1-24

[72] Godínez ML, Tovar-Sánchez E, Valencia-Cuevas L, Rosas-Ramírez ME, Rodríguez A, Mussali-Galante P. Glyphosate pollution treatment and microbial degradation alternatives, A review. Microorganisms. 2021;**9**(11):2322

[73] Takkar S, Shandilya C, Agrahari R, Chaurasia A, Vishwakarma K, Mohapatra S, et al. Green technology: Phytoremediation for pesticide pollution. In: Phytoremediation Technology for the Removal of Heavy Metals and Other Contaminants from Soil and Water. Elsevier; 2022. pp. 353-375

[74] Mishra SK, Kumar PR, Singh RK. Transgenic plants in phytoremediation of organic pollutants. In: Bioremediation of Pollutants. Elsevier; 2020. pp. 39-56

[75] Saeed Q, Xiukang W, Haider FU, Kučerik J, Mumtaz MZ, Holatko J, et al. Rhizosphere bacteria in plant growth promotion, biocontrol, and bioremediation of contaminated sites: A comprehensive review of effects and mechanisms. International Journal of Molecular Sciences. 2021;**22**(19):10529

[76] Das PK, Mohanty C, Purohit GK, Mishra S, Palo S. Nanoparticle assisted environmental remediation: Applications, toxicological implications and recommendations for a sustainable environment. Environmental Nanotechnology, Monitoring & Management. 2022;**18**:100679

[77] Susaimanickam A, Selvaraj D, Manickam P. Nanoparticle-based bioremediation of organic and inorganic substances from water bodies. In: Bioremediation of Emerging Contaminants in Water. Vol. 2. American Chemical Society; 2024. pp. 135-162

[78] Raj A, Kumar A, Dames JF. Tapping the role of microbial biosurfactants in pesticide remediation: An eco-friendly approach for environmental sustainability. Frontiers in Microbiology. 2021;**12**:791723

[79] Halecký M, Kozliak E. Modern bioremediation approaches: Use of biosurfactants, emulsifiers, enzymes, biopesticides, GMOs. In: Advanced Nano-Bio Technologies for Water and Soil Treatment. 2020. pp. 495-526

[80] Lopes PRM, Cruz VH, De Menezes AB, Gadanhoto BP,

Moreira BRDA, Mendes CR, et al. Microbial bioremediation of pesticides in agricultural soils: An integrative review on natural attenuation, bioaugmentation and biostimulation. Reviews in Environmental Science and Bio/Technology. 2022;**21**(4):851-876

[81] Kamal B, Goswami R, Mishra A. The environmental implication and microbial remediation of pesticide pollution: A critical assessment of the concept, strategies, and future perspective. In: Pesticides Bioremediation. Cham: Springer International Publishing; 2022. pp. 165-193

[82] Mishra S, Lin Z, Pang S, Zhang W, Bhatt P, Chen S. Recent advanced technologies for the characterization of xenobiotic-degrading microorganisms and microbial communities. Frontiers in Bioengineering and Biotechnology. 2021;**9**:632059

[83] Rajan R, Wani AW, Mirza A, Sultana S, Singh J. Genome editing of microbes for degradation of pesticides. In: Current Developments in Biotechnology and Bioengineering. Elsevier; 2023. pp. 167-200

[84] Fuller E, Castro-Gutiérrez V, Cambronero-Heinrichs JC, Rodríguez-Rodríguez CE. Using molecular methods to identify and monitor xenobiotic-degrading genes for bioremediation. In: Biodegradation, Pollutants and Bioremediation Principles. 2021. pp. 65-90

[85] Singh J, Gupta M, Singh KK, Kumar A, Yadav D, Wenjing W, et al. Advancement in bioinformatics and microarray-based technologies for genome sequence analysis and its application in bioremediation of soil and water pollutants. In: Microbe Mediated Remediation of Environmental Contaminants. Woodhead Publishing; 2021. pp. 209-225

Chapter 5

Mycoremediation: An Innovative and Sustainable Approach

Dalel Daâssi, Mouna Bouassida, Fatimah Almaghrabi and Mohamed Chamkha

Abstract

The expanding population, rapid growth in urbanization, and industry exacerbate pollution-related issues. Pollution of the soil and water has an impact on both human health and the environment. Thus, the restoration and the cleanup of contaminated areas represent an important technological and environmental challenge for sustainable growth and development. Among the various techniques used to remediate environmental contaminants, Mycoremediation, the use of fungi or its derivatives, is by far the most widely employed to remove or degrade metal metals, persistent organic pollutants, and some emerging pollutants. This chapter summarizes the key aspects of mycoremediation as an eco-friendly, economical, and sustainable approach to environmental remediation in both terrestrial and aquatic ecosystems. The current chapter discusses the potential of various fungi in degrading contaminants such as heavy metals, agricultural and pharmaceutical wastes, dyes, and polycyclic aromatic hydrocarbons. Likewise, we address the major fungal species, their mechanisms, and applications for mycoremediation. Afterward, the economic and environmental benefits, problems, and future techniques for improving the efficiency of remediation are addressed.

Keywords: mycoremediation, polycyclic aromatic hydrocarbons, heavy metals; pharmaceutical wastes, dyes, pesticides

1. Introduction

Pollution is a significant concern. The insufficient management of waste and effluents from households, industries, and agricultural fields is further damaging the ecosystem. According to World Health Organization investigations, 2.2 billion people do not have access to safe water services, while 144 million people use contaminated water. Half of the world's population is estimated to reside in water-stressed areas by 2025 [1].

Currently, xenobiotic contaminants may be removed and degraded from soil and water using a variety of physical and chemical processes. However, these procedures are costly, produce unsafe byproducts, and are inefficient for low concentrations of highly toxic compounds [2]. A solution must be created to overcome these limitations and make pollutant removal possible.

Bioremediation has become a promising biological treatment for restoring contaminated areas. It has been established as efficient, economical, versatile, and environmentally eco-friendly [3]. Bioremediation uses living organisms selected based on their metabolic diversity and performance to remove or reduce contaminant levels [4].

Fungi-mediated bioremediation, or mycoremediation, has been studied in both terrestrial and aquatic habitats. In recent years, detoxification of contaminated sites by the application of one or more species of fungi as natural agents has become a widely used and effective tool for environmental remediation. It has become practical, economical, and efficient in converting toxic wastes into not/or less toxic end products or dioxide carbon and water [5]. Compared to other microorganisms, fungi have strong morphology, extensive hyphal network, adaptation to extreme circumstances, and tolerance for elevated amounts of pollution [6]. Similarly, fungi swiftly absorb contaminants and may survive in conditions that have inadequate nutrient contents, acidic pH, or low water activity (WA) [7]. Furthermore, fungal species may produce versatile extracellular enzymes that can interact with diverse xenobiotic structures with a high degree of non-specificity. So, their diversity and the ability to release a wide range of specific and non-specific enzymes make fungi ideal candidates for addressing a wide spectrum of pollutant materials [8]. It can be efficient in remediating, in situ or ex-situ, several pollutants such as heavy metals, dyes, herbicides, and pharmaceuticals generated from various industries [9–13].

Alternatively, Mycoremediation implicates different approaches like fungal augmentation (biomass addition) or biostimulation (nutriment supplementation) to enhance pollutant removal [14]. Those approaches can be applied in bioreactors with regulated physicochemical conditions designed to promote simultaneously microbial growth and degradation rates [15]. A wide variety of fungal species have shown their potential to degrade a wide range of environmental pollutants. The most common fungal strain which has been recorded as a biodegrade belongs to the following genera: *Cephalosporium, Rhizopus, Paecilomyces, Fusarium, Emericella, Trichoderma, Torulopsis, Mucor, Talaromyces, Gliocladium, Rhodotorula, Cladosporium, Geotrichum, and Aspergillus* [15].

2. Fungi as bioremediating agents

Fungi typically play important roles in ecosystems as endophytes, infectious agents, and saprophytes. The current estimate of fungal diversity is quite uncertain, ranging from 1.5 to 12 million, with only 1 million being described. Fungi are classified as eukaryotes. The most recent taxonomy update covers 19 major phyla [16].

Fungi are the most effective decomposers of organic matter. They are involved in the delignification of plant-based materials such as lignin or cellulose [17]. Unlike other microbes, fungi are resistant to destruction because their cell walls are made of chitin and melanin.

Fungi secrete a range of enzymes and organic acids which can target and decompose many inorganic and organic contaminants [18]. Fungal enzymes are frequently characterized by low specificity against substrate [19]. Contrary to popular belief, this gives fungi an advantage since non-specific enzymes catalyze a wide range of reactions, allowing them to employ a variety of chemicals as carbon and energy sources. Fungi's biodegradation abilities are usually associated with oxidative enzymatic activities mediated by a wide variety of oxidases and peroxidases [20]. As a result, fungus often requires an oxygen-rich environment to function properly.

Oxidoreductases are the most studied extracellular fungal enzymes, including laccases, which require oxygen, and peroxidases (manganese peroxidase, lignin peroxidase), which apply H_2O_2 as terminal electron acceptor [19, 20]. One of the most well-known extracellular pollutant-degrading processes is hydroxyl radical oxidation, which cleaves double bonds in cyclic or aliphatic structures. The hydroxyl

Pollutants	Fungi	Mechanism	References
Polycyclic aromatic hydrocarbon	*Aspergillus niger (MW699896), Paecilomyces formosus(MW699897), Fusarium chlamydosporum (MZ817957); Coniochaeta sp. (MW699893);Coriolopsis gallica (KJ 412304); Trametes versicolor; Dentipellis sp. (KUC8613); Phanerochaete chrysosporium; Pleurotus ostreatus; Pleurotus eryngii, Cochliobolus lunatus, Trichoderma longibrachiatum*	Cytochrome P-450 monooxygenase, dioxygenase, dehydrogenases, glutathione, FAD-dependent monooxygenases, Ligninolytic enzymes, transferase, and epoxide hydrolases mediated degradation	[23–26]
Heavy metals	*White-rot fungi (Phlebia brevispora and Phlebia foridensis), Aspergillus species, Cladosporium sp. NRCA8 Trichoderma harzianum, Trichoderma ghanense, Rhizomucor species, Penicillium rubens, Sordariomycetes sp. (D10), and Coniochaetaceae sp. Fusarium* species, *Emericella species, Funneliformis geosporum, Pleurotus ostreatus,*	Ligninolytic enzymes in the degradation of heavy metals, and antioxidant enzymes in tolerating damage due to oxidative stress	[10, 27]
Dyes	*Aspergillus flavus, Marasmius cladophyllus, Phlebia acerina, Bjerkandera adusta, Aspergillus niger, and Fusarium proliferatum*	Laccase, Manganese peroxidase, and Lignin peroxidase in degradation of dyes	[28–30]
Pesticide and herbicide	*Botryosphaeria laricin*a, *Aspergillus glaucus, Trametes pavonia, Penicillium spiculisporus, Penicillium verruculosum,*	Ligninolytic enzymes, esterification, dioxygenation, dehydrogenation, dechlorination, demethylation mediated degradation	[12, 31, 32]
Antibiotics	*Pleurotus ostreatus, Leptosphaerulina sp, Irpex lacteus, Lentinula edodes*	versatile peroxidase, laccase, manganese peroxidase, cytochrome 450 system	[13, 33]
Pharmaceuticals	*Mucor hiemalis, Trametes versicolor, Phanerochaete chrysosporium, Lentinula edodes,*	ligninolytic enzymes and cytochrome 450	[34, 35]
Phthalates	*Fusarium oxysporum, Fusarium subglutinans, Penicillium brocae, Purpureocillium lilacinum*	Cutinase	[36]
Detergents	*Penicillium verrucosum, Cladsporium cladosporioides,* and *Geotrichum candidum*	NA[(a)]	[37]

(a) Not applicable.

Table 1.
Pollutants and its degradation by fungal species.

radicals are the result of quinone cycling following laccase action [21] or peroxide-dependent hydroxylations. Both mechanisms are exergonic in nature, and they aid in the breakdown of complicated pollutant structures into more easily degradable metabolite intermediates. Likewise, Cytochromes catalyze a wide range of oxidation events within cells, with some species, such as *Phanerochaete chrysosporium,* including 150 Cytochrome P450 genes in their genome [22]. In the scientific literature, several studies have reported the potent of different fungi to accumulate and remove various xenobiotic pollutants from the environment (**Table 1**).

3. Mechanisms of mycoremediation

The environment contains a huge number of extremely complex, hazardous, and persistent xenobiotic and/or recalcitrant chemicals resulting from numerous human activities [5]. Examples of these activities include chemical fertilizers and pesticides in agriculture, antibiotics, food additives, and personal care items. Mining, petrochemical, food, and textile industries generate waste and effluent containing heavy metals, PAHs, dyes, phenols, and other recalcitrant compounds, which can contaminate soil and water bodies if not properly treated before discharge [5]. Mycoremediation is a promising technology for environmental cleanup due to its cost-effectiveness, environmental friendliness, and versatility. Fungi can be used to remediate a wide range of pollutants in various environments, using physical or enzymatic mechanisms such as biosorption and/or bioaccumulation, biodegradation, and bioconversion [38].

3.1 Biosorption and/or bioaccumulation

Generally, biosorption is the ability of certain proteins or microbial biomass (living or dead) to bind and concentrate ions or other molecules [11, 39]. Thus, the basic idea of bioadsorption, which is a passive phenomenon carried out by dead biomass or specific molecules, is based on the "affinity" between the bio-adsorbent and the sorbate, which corresponds to the fungal biomass and the contaminant. This binding can occur through various mechanisms, including ion exchange, complexation, and physical adsorption. Indeed, the negative charge of several functional groups present in fungal cell walls, including proteins, lipopolysaccharides, and phospholipids, enables them to effectively bind metal cations [40]. Numerous benefits, including low nutrient requirements, high efficiency, little sludge creation, and the potential for metal regeneration and recovery, make biosorption by fungal biomass an economically appealing and technically possible method [11].

Biosorption is particularly effective for heavy metals, which can bind strongly to the negatively charged cell wall of fungi (**Figure 1**) [41].

Several studies have shown that fungi may remove various substances through mycelial biosorption, biomineralization, and biotransformation [42, 43]. Fungal biomass, whether alive or dead, can effectively absorb various pollutants due to its diverse functional groups in the cell wall [44]. In the literature, heavy metals, herbicides, dyes, and drugs have been efficiently absorbed by mycelial pellets more than mycelium [45, 46]. Fungal biomass, both living and dead, can act as absorbents to remove heavy metal ions. Due to its low cost, the potential for toxicant recovery, the ability to regenerate and reuse dead biomass, and simple maintenance, dead fungal biomass is thought to be superior to living fungi [46]. It is hypothesized that fungi are able to metabolize the substances they have obtained by biosorption, in some cases even

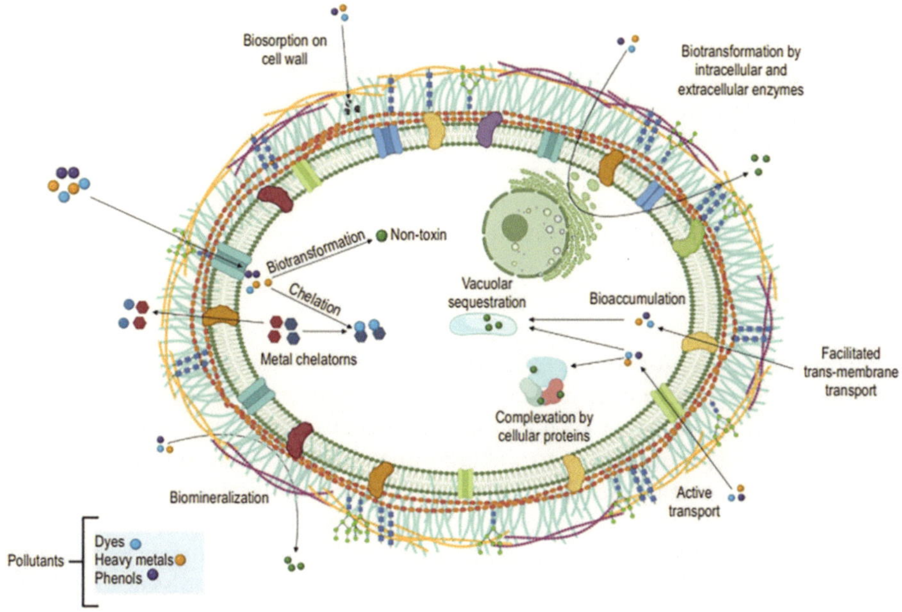

Figure 1.
Mycoremediation mechanisms for removal of inorganic and organic pollutants (Adapted from Shourie and Vijayalakshmi [41]).

mineralizing them, in a process of bioconversion and biodegradation that has been investigated in studies on cleaning the environmental matrix of various pollutants.

Generally, pollutant biosorption occurs through two known processes. The first, referred to as the passive mode, involves the uptake of substances independent of metabolic activity. During this process, mechanisms such as precipitation, physical adsorption, ion exchange, and complexation lead to the binding of pollutants like metals, to the surface of cell walls and extracellular material [47]. These pollutants are adsorbed onto the fungal cell surface, decreasing their mobility and bioavailability in the environment. The second mode, which involves moving metals across the cell membrane, is an effective, active, and energy-dependent process [48].

3.2 Biodegradation

Biodegradation involves the enzymatic breakdown of pollutants by fungi. Fungi have been found to produce a wide range of extracellular enzymes that can degrade various organic and inorganic compounds, including ligninolytic enzymes [49] and non-ligninolytic enzymes [50] that convert persistent organic and inorganic pollutants into less toxic smaller units that are utilized for their growth. Further, in some cases, biodegradation can lead to complete mineralization of the pollutant, converting it into harmless substances like carbon dioxide and water.

Ligninolytic enzymes, such as lignin peroxidase and manganese peroxidase, are produced by white-rot fungi and can degrade a wide range of recalcitrant pollutants, including polycyclic aromatic hydrocarbons (PAHs), pesticides, and dyes [12, 23, 30, 31].

Typically, Oxidoreductases, including laccases, are the most common enzymes in the biodegradation process of several environmental pollutants, due to their versatility and active center that can accommodate a wide range of substrates [51].

Laccases have the extra advantage of requiring only molecular oxygen to catalyze processes. Furthermore, their production is ubiquitous across organisms such as bacteria, fungi, and plants, with white-rot fungi (WRF) being particularly efficient producers [12, 30, 31].

Additionally, fungal laccases have high activity and interesting qualities that are important in bioremediation processes, such as substrate specificity, pH, and the enzyme's optimum temperature, which are also used in wastewater treatment plants (WWTPs) for pollutant removal [51].

3.3 Bioconversion

Bioconversion is a process where fungi transform pollutants into new compounds, which may be less toxic or even useful. For example, some fungi can convert toxic metals into less toxic forms or even accumulate them in their biomass. Bioconversion can also lead to the production of valuable compounds, such as enzymes, antibiotics, and biofuels [52]. Several research studies have shown that ligninolytic enzymes have a part in the breakdown of certain xenobiotic pollutants.

Likewise, Various fungal species possess the ability to biotransform pollutants through processes such as reduction, oxidation, methylation, catalysis, and dealkylation [48].

4. Applications of mycoremediation

Pollutants generated from intensive industrialization and various human activities, including waste and effluents from households, industries, and agricultural sectors, pose significant risks to human health and degrade the ecosystem [53].

The versatility of fungi enables their application across these various environmental contexts for remediation. This part summarizes the roles of various fungi in the degradation of recalcitrant, persistent, and harmful pollutants such as polycyclic aromatic hydrocarbons, heavy metals, agricultural pollutants including pesticides, herbicides, and insecticides, also pharmaceuticals like drugs and antibiotics.

4.1 Polycyclic aromatic hydrocarbons (PAH) and their remediation

Fungal hydrocarbon degradation occurs through two pathways: a non-enzymatic process which involves a series of biochemical reactions, including biosorption, biomineralization, and stripping, and an enzymatic process, which is further divided into ligninolytic and non-ligninolytic enzymes.

During the non-ligninolytic enzyme process, enzymes like those from the cytochrome P450 family activate hydrocarbon mixtures through oxidation, reduction, hydrolysis, and dehalogenation processes. These activated hydrocarbon compounds are transformed into intermediate metabolites before being converted to excreted derivatives [54]. Similar studies have shown the degradative mechanism of PAHs. For instance, it was reported that *Dentipellis sp.* KUC8613 first utilizes cytochrome P450, followed by enzymes such as dioxygenase, dehydrogenases, FAD-dependent monooxygenases, glutathione transferase, and epoxide hydrolases to degrade PAH [55]. Other studies demonstrated how ligninolytic enzymes, such as laccase, lignin peroxidase, and manganese peroxidase, may break down petroleum hydrocarbons because of their structural resemblance to lignin [55]. These extracellular ligninolytic enzymes catalyze the oxidation of PAH into quinones, which are then degraded through ring fission [54]. Fungal species such as *Trametes spp.*, *Lentinus spp.*, *Pleurotus spp.*, and *Ganoderma spp*. are highly effective in PAH biodegradation due to their ability to produce non-specific enzymes such as laccase, manganese peroxidase, and lignin peroxidase [26].

Yet, other studies showed that petroleum hydrocarbons could activate the enzymatic system. Azin et al. [56] reported that the laccase activity of *F. neocosmosporiellum RH-10* reached its maximum level after 20 and 10 days in the absence and presence of 10 g/L crude oil, respectively.

Afterward, several studies demonstrated that fungal consortia can enhance the rate of petroleum pollutant degradation. Similarly, Nrior and Onwuka [57] studied the biodegradation of marshland crude oil-contaminated soil through bioaugmentation using both individual and combined species of *Candida* and *Penicillium*. They found that the fungal consortium had higher bioremediation capacity than the individual species. Additionally, *Candida* species demonstrated better bioremediation capabilities compared to Penicillium species, owing to their effective degradative enzyme system, which enables them to break down both long-chain and short-chain hydrocarbons.

Basically, three main approaches have been used for petro-hydrocarbon fungal bioremediation including bioaugmentation, biostimulation, and a combination of both [58]. For instance, the bioaugmentation approach has been especially effective, reducing BaP to 25.75 mg/kg in 28 days and Phe to 0.4 mg/kg in 21 days [59].

Additionally, periodic biostimulation and bioaugmentation with a fungal consortium consisting of *Pestalotiopsis sp.*, *Polyporus sp.*, and *Trametes hirsuta* were found to be effective biodegradation methods in a petroleum hydrocarbon bioremediation study [60]. Other studies reported that associating PAHs with plants presented another approach to their removal from soil. For instance, in the presence of *Zea mays*, *Pleurotus ostreatus* facilitated the mycoremediation of PAHs, enhancing fungal, microbial, and manganese peroxidase activity [61].

Typically, WRFs, such as *P. chrysosporium* and *Trametes versicolor*, are reported to be good BTEX degraders together with soil and mycorrhizal fungi [62]. Numerous studies have shown a direct correlation between crude oil degradation and surfactant synthesis. According to Bhardwaj et al. [62], the fungal biosurfactants have diverse applications in the oil and gas industry, including enhancing oil recovery from reservoirs, remediating oil spills, and decontaminating soil and water polluted with hydrocarbons.

4.2 Heavy metal pollution and its mycoremediation

Generally, heavy metals can originate from natural processes, such as volcanic emissions, continental dust transport, and metal-rich rock weathering. They are also generated through anthropogenic activities including, mining, smelting, fossil fuel combustion, waste disposal as well as agricultural, industrial, and military practices [63]. Cd, Cr, Hg, Pb, Cu, Zn, and As are the most common heavy metals found in soil [26]. Many of these metals are known for their mutagenic and carcinogenic effects. In the case of microorganisms, enzyme activities such as urease, catalase, glucosidase, and alkaline phosphatase can be affected by these metals [64]. In plants, the accumulation of these metals may hinder various physiological activities including mineral and water transport and photosynthesis [5]. Because of their properties, fungi are a very effective alternative for the elimination of heavy metals. They are more environmentally adapted and exhibit higher tolerance to heavy metals than bacteria and algae [5].

For instance, *Pleurotus pulmonarius* was used at different concentrations for 62 days to study the mycoremediation of soil contaminated by petroleum hydrocarbons. The amounts of heavy metals (zinc, copper, and manganese) and hydrocarbons were found to have significantly decreased. At the 10% mycelium concentration, the greatest decreases in Mn (24.00 to 1.7), Cu (37.24 to 0.00), and Zn (63.03 to 5.75) were observed by Njoku et al. [26].

Additionally, when inoculated in Pb-contaminated soil microcosms, an isolate of *Chaetomium aureum* was able to reduce the concentration of free Pb in the soil by half in around 2 months [65].

Metal bioprecipitation, biosorption, biomineralization, biotransformation, and bioleaching are examples of extracellular processes. However, exposure to heavy metals causes intracellular bioaccumulation, activating a variety of signaling pathways and metabolic processes aimed at maintaining cellular homeostasis [28]. For instance, *Aspergillus*, *Fusarium*, *Rhizomucor,* and *Emericella spp.,* isolated from highly contaminated agricultural soil, showed different detoxification mechanisms (biosorption/bioaccumulation and biovolatilization).

Biosorption, a passive process which can be carried out by both live cells and dead biomass [5], is one of the main ways that fungi remove heavy metals from contaminated environments [14]. The fungal cell surface serves as the primary defense against metals and metalloids, containing wall polysaccharides, lipids, proteins, and functional groups like carboxylate, hydroxyl, amino, and phosphate that can bind heavy metal ions. Moreover, chitin and chitosan, which are structural components of fungal cell walls, have shown effectiveness in metal binding across various fungal species [66].

Another active method of removing metal is called bioaccumulation, which uses an energy-dependent metal inflow mechanism. Compared to biosorption, it is a slower process [52]. Fungi, in particular basidiomycetes mushrooms, are potent heavy metal accumulators. They absorb metals from soil through their mycelia and accumulate them in their fruiting bodies [27].

For instance, *Trichoderma harzianum* can bioaccumulate significant amounts of copper and zinc, effectively removing these metals from contaminated soil [27]. In another study, this fungus growing in areas around gold, silver, and gemstones mine was found to accumulate these compounds, offering an economic advantage [5].

Several fungal species possess the ability to biotransform metals through processes such as reduction, oxidation, methylation, catalysis, and dealkylation (**Table 1**) [52].

These processes convert hazardous metals and metalloids into less toxic forms that reduce the concentrations of heavy metals in the environment [67]. Enzyme secretion and mechanical assistance of fungal hyphae enhance the transformation of pollutants [67].

Alternatively, intracellular enzymatic reactions may convert inorganic and organic compounds into their volatile derivatives through biovolatilization [5]. The enzymatic process can transfer metals like Hg, Sn, and Pb to the methyl group, resulting in molecules with varied solubility, volatility, and toxicity [67].

Furthermore, redox processes play a critical role in the transformation and detoxification of contaminants by fungi. During these processes, the pollutant is reduced or oxidized through electron transfer between the pollutant and the fungal cells. Similarly, Sepehr et al. [68] reported that in both aerobic and anaerobic settings, *Aspergillus niger* has been shown to convert Cr(VI) to Cr(III). The *Aspergillus* species have been demonstrated to be a powerful bioremediation agent for chromium-contaminated environments.

Other recent studies reported that fungi can induce precipitation reactions by converting soluble metal ions into insoluble forms that can be separated by sedimentation or filtration [14]. In this context, *Aspergillus niger* has been reported to exhibit good potential in generating a variety of organic acids like citric and oxalic acids effective for precipitating heavy metals from contaminated soil. Citric acid can form stable complexes with metals such as lead, cadmium, and aluminum, decreasing their toxicity and bioavailability [14].

4.3 Pharmaceuticals "fungal remediation"

According to reports, natural streams and rivers have been found to contain pharmaceutical chemicals in a range of concentrations that can be harmful to aquatic biota [69]. Commonly utilized water treatment systems fail to eliminate these xenobiotics, resulting in the accumulation of adverse pharmaceutical chemicals in the environment [70].

Another common issue is antimicrobial resistance (AMR). These antibiotics may harm ecological microbes that are involved in many natural processes such as carbon cycling, soil respiration, iron reduction, denitrification, and nitrification [5].

Mycoremediation is a cost-effective, environmentally friendly, and safer alternative to traditional physical and chemical approaches for treating antibiotic-contaminated water.

Several fungi have been observed to remove different antibiotics such as bifonazole, clotrimazole, sulfonamides, oxacillin, oxytetracycline, and fluoroquinolone [71].

Recent studies reported the potential of both aquatic and white-rot fungi (WRF) as promising candidates for remediating pharmaceutical discarded residues. According to Nosek and Zhao [72], the use of WRF or ligninolytic enzymes has shown encouraging results, with degradation yields for diclofenac (DCF) approaching 100%. These degradation systems relied on the utilization of laccase as the biocatalyst, which could be introduced exogenously together with an appropriate mediator, 1-hydroxibenzotriazole (HBT), or endogenously produced in batch fungal cultures for in vivo degradation. However, the function of ligninolytic enzymes, such as versatile peroxidase and laccase, has been documented in the breakdown of several isoxazolyl penicillins such as oxacillin and dicloxacillin by *Leptosphaerulina sp.* in water [73].

In some studies, aquatic fungi, like *Mucor hiemalis,* have been reported as efficient for removing acetaminophen from contaminated water bodies caused by pharmaceutical effluent disposal [70].

Similarly, the potential of immobilized *Trametes versicolor* for the remediation of pharmaceutical chemicals from urban wastewater was also shown by Del Álamo et al. [74].

4.4 Agricultural waste and its mycoremediation

Herbicides and pesticides have assisted farmers around the world in combating numerous plant and insect illnesses and enhancing agricultural productivity. On the contrary, these substances have a negative impact on people and the environment. Presently, most pesticides and herbicides wash off to adjacent water sources or deposit in the soil and enter the food chain, affecting microbes, plants, animals, and humans [75].

Mycoremediation appears to be a viable, environmentally friendly, cost-effective, and on-site technique for treating herbicide—and pesticide-contaminated soil and water. In this context, the study of Matúš et al. [76] reported the performance of *Aspergillus* and *Penicillium* species for the biodegradation of organochlorine and organophosphorus pesticides. In an additional study, it was found that a unique strain of *Aspergillus glaucus was* able to metabolize fipronil and its metabolite, fipronil sulfone using ligninolytic enzymes and biotransformation processes [77].

According to Chan-Cupul et al. [66], *Trametes maxima* enzyme preparations, as well as a co-culture of *T. maxima* and *Paecilomyces carneus*, have been shown to completely remove atrazine.

A few studies have shown esterification, deoxygenation, dehydrogenation, dechlorination, and demethylation reactions during the fungi-mediated transformation of these herbicides [78]. Yet another xenobiotic chemical, glyphosate, is effectively digested by fungus-like *Penicillium spiculisporus*, *Aspergillus flavus*, and *Penicillium verruculosum* [79].

4.5 Dyes

Large amounts of wastewater containing various dyes are released into the environment by the printing, textile, and dye-producing industries. These dyes' intrinsic qualities of being chemically stable, resistant to fading, and immune to light and microbial degradation make them stubborn and long-lasting in the environment. Most dyes are not readily biodegradable by natural microorganisms, hindering their natural removal from the environment [5].

Bioremediation techniques are far superior to conventional treatments for the degradation of dye. In fact, a wide range of fungal strains is capable of removing a diverse range of dyes. They can be used to eliminate textile dyes via biotransformation, biodegradation, and/or adsorption by the biomass [48]. It was recently shown that a strain of *Penicillium oxalicum* has the ability to biodegrade the dyes Direct Red 75, Acid Red 183, and Direct Blue 15 [80]. *Aspergillus niger* can decolorize a red azo dye (1000 mg/L dye concentration) at pH 9.0 [81].

In addition, the endophytic fungi Marasmius cladophyllus, isolated from *Melastoma malabathricum*, was also able to degrade different classes of synthetic dyes [82]. Moreover, endophytic fungi are used for detoxification of Crystal Violet (CV), Malachite Green (MG), Methyl Violet (MV), and cotton blue found in industrial effluents [83].

Multiple studies investigating the effectiveness of using fungi to clean dye wastewater concluded that growing fungi, immobilized fungi, entire granular cell components, and mycelial pellets displayed good removal efficiency [84].

As an example, Przystaś et al. [85] reported the efficiency of immobilized biomass of Pleurotus ostreatus, Gloeophyllum odoratum, and Polyporus picipes in decoloring Evans Blue and Brilliant Green, azo and triphenylmethanol dyes, respectively.

In recent years, mycelial pellet technology for decolorizing dye effluent has been studied. Fungal pellets enhance rheology by reducing viscosity and enhancing oxygen delivery to the biomass when they are introduced to bioreactors for wastewater treatment [86].

5. Challenges and future directions

Despite the advantages of mycoremediation, challenges remain for its industrial-scale application. One major challenge is the variability in efficacy among different fungal genera when dealing with toxicants. The degradation rates and activity levels of fungi may vary based on the type of contaminants present in the environment [87].

Understanding the complex interactions between fungi and contaminants is essential as this variability necessitates the development of tailored approaches to select and apply appropriate fungal strains for specific pollutants.

Furthermore, the ecological interactions within soil ecosystems remain insufficiently understood which can affect remediation outcomes. Exploring fungal adaptability across ecosystems can aid in the development of tailored mycoremediation strategies for diverse contamination scenarios.

In some cases, WRF inhibits soil bacterial growth, whereas in other cases, soil bacteria inhibit WRF proliferation. The toxicity of contaminants can also affect fungal viability and soil biota, necessitating pollutant monitoring to ensure successful degradation [23].

Another significant challenge lies in enzymatic activity. For instance, soil physiochemical factors including moisture, carbon-to-nitrogen ratios, oxygen levels, temperature, and pH, play a significant role in influencing fungal enzyme production and activity. These factors are more variable in contaminated soils compared to controlled environments [23]. Some studies in the scientific literature have shown that degradation can be enhanced through the co-addition of fungi and isolated enzymes, such as laccase. Similarly, Fan et al. [88] reported that co-remediation using WRF and laccase extract accelerates and enhances the degradation of 2,2-Bis(p-chlorophenyl)-1,1,1-trichloroethane in the soil matrix more effectively than employing WRF or laccase alone. Further studies are needed to explore how enzyme manipulation and addition can optimize degradation efficiency while evaluating the economic viability of these approaches.

To address these challenges, several enhancement strategies are emerging. Genetic modification offers a promising pathway for improving fungi resilience against toxic pollutants. With rapid advancement in genetic engineering and synthetic biology, future research should focus on using these technologies to improve the performance of fungi-based treatment systems [89]. For instance, genetic engineering can improve the fungal tolerance to metals and metalloids by increasing extracellular and intracellular metal chelators, modifying metal transporters, overproducing antioxidative enzymes, and altering regulatory networks [89]. Post-genomic approaches provide new insights into fungal gene functions, enabling a better understanding of cell responses to metal toxicity [71]. However, public acceptance of genetically modified organisms (GMOs) must be considered.

Nanotechnology also offers novel opportunities to enhance mycoremediation. The integration of fungi with nanoparticles (NPs) has been shown to improve pollutant adsorption, catalysis, and degradation. Fungi possess several advantages over other microorganisms for NPs synthesis including ease of handling, simple nutrients requirements, high cell wall-binding capacity, and significant intracellular metal uptake [71].

The synergy between fungal biosynthesis and nanomaterial development has gained increasing attention. For example, Velmurugan et al. [90] demonstrated that Fusarium sp. from Zn-contaminated sites could bioabsorb up to 320 mg/l Zn and produce ZnO nanoparticles.

Similarly, immobilized *P. chrysosporium* loaded with TiO2 nanoparticles has been shown to be novel high-value bioremediation materials capable of adsorbing Cd and degrading 2,4-dichlorophenol where the antioxidative defense system and the physiological fluxes protect the system from oxidative stress. Zuo et al. [91] reported a tenfold increase in Cd(II) removal by Phanerochaete chrysosporium in the presence of Ag nanoparticles.

Interestingly, fungi have also been effective in removing NPs from an aqueous medium. Jakubiak et al. [92] found that *Pleurotus eryngii* and *Trametes versicolor* could efficiently remove Al_2O_3 NPs, up to 86 and 61% of the total amount of NPs, respectively. Additionally, killed biomass of *Hypocrea lixii* has been successfully used to convert Cu and Ni ions into CuO and NiO NPs in aqueous solution [93].

Yet, another challenge was reported for the nanotechnology-based mycoremediation. Recently, in the scientific literature, it was demonstrated that this approach of remediation, based on nanotechnology techniques, reduces air pollution through contaminant adsorption using nanoabsorptive materials, degradation via nanocatalysis, and filtration/separation using nanofillers [94].

Further, substrate and encapsulation techniques represent additional strategies to enhance WRF colonization and pollutant degradation. Substrates are organic materials combined with fungi to serve as carbon and energy sources before being introducing into the soil. Substrates such as wood chips, plant waste, and fish oil can enhance fungal growth, but the most effective substrate remains unidentified [95]. Substrates can be directly mixed into contaminated soil or encapsulated before soil inoculation. Encapsulates are made by mixing chemicals, substrates, nutrients, and possibly enzymes into casings that enclose fungal spores.

Although fungi are powerful in degrading various pollutants at the laboratory scale, and/or reactors, in situ remediation remains challenging. The scaleup requires optimization of processes such as optimization of the reactor design, selecting appropriate substrate, and adjusting operational parameters as well as conducting economic feasibility studies. It was reported that a phased approach involving bench-scale testing, pilot studies, inoculum production, and full-scale application is recommended [96].

Furthermore, future progress will be guided by a discussion on the financial requirements for full-scale implementation. Overcoming financial obstacles will require cooperation with governments and businesses to support research and development. Lastly, public opinion and legal frameworks are important factors in the uptake of fungal technology. Initiatives for education and public outreach can contribute to the development of mycoremediation technologies' credibility and support [96]. To further reassure the public, it is also essential to validate the long-term efficacy of soil remediation and demonstrate the safety of byproducts.

6. Conclusions

Mycoremediation, the use of fungi to degrade pollutants, holds immense promise for environmental cleanup. Fungi possess remarkable metabolic capabilities, enabling them to degrade a wide range of persistent and toxic pollutants, including polyaromatic hydrocarbons, heavy metals, pharmaceuticals, and agricultural wastes. Studies show the role of extracellular ligninolytic enzymes and cytochrome 450 in degradation. However, the underlying mechanism is elusive and requires further research. Genetically improved fungi can be used for efficient and rapid pollution removal, and bioreactor conditions should be optimized.

Acknowledgements

This project was funded by The Ministry of Higher Education and Scientific Research (MESRS) through the Programme for Encouraging Young Researchers

(PEJC) session 2023, Tunisia, under Grant No. (PEJC2023-D1P02). Therefore, the authors acknowledge and thank the MESRS for its technical and financial support.

Conflict of interest

The authors declare no conflict of interest.

Author details

Dalel Daâssi[1*†], Mouna Bouassida[1†], Fatimah Almaghrabi[2] and Mohamed Chamkha[1]

1 Laboratory of Environmental Bioprocesses, Centre of Biotechnology of Sfax, University of Sfax, Sfax, Tunisia

2 The Applied-College Khulais, University of Jeddah, Jeddah, Saudi Arabia

*Address all correspondence to: dalel.daassi@cbs.rnrt.tn

† These authors contributed equally.

References

[1] World Health Organization: WHO. Drinking-Water. 2023. Available from: https://www.who.int/news-room/fact-sheets/detail/drinking-water.

[2] Crini G, Lichtfouse E, Wilson LD, Morin-Crini N. Conventional and non-conventional adsorbents for wastewater treatment. Environmental Chemistry Letters. 2018;**17**(1):195-213. DOI: 10.1007/s10311-018-0786-8

[3] Kuppan N, Padman M, Mahadeva M, Srinivasan S, Devarajan R. A comprehensive review of sustainable bioremediation techniques: Eco friendly solutions for waste and pollution management. Waste Management Bulletin. 2024;**2**(3):154-171. DOI: 10.1016/j.wmb.2024.07.005

[4] Dangi AK, Sharma B, Hill RT, Shukla P. Bioremediation through microbes: Systems biology and metabolic engineering approach. Critical Reviews in Biotechnology. 2018;**39**(1):79-98. DOI: 10.1080/07388551.2018.1500997

[5] Akhtar N, Mannan MA. Mycoremediation: Expunging environmental pollutants. Biotechnology Reports. 2020;**26**:00452. DOI: 10.1016/j.btre.2020.e00452

[6] Gong Z, Zhang S, Liu J. Recent advances in chitin biosynthesis associated with the morphology and secondary metabolite synthesis of filamentous fungi in submerged fermentation. Journal of Fungi. 2023;**9**(2):205. DOI: 10.3390/jof9020205

[7] Che Y, Wang J, Ding W, Xiao Z, Shi X, Wu J, et al. New diketopiperazine alkaloids from the Haima cold seep-derived fungus Toxicocladosporium sp. CYH-18. Phytochemistry Letters. 2024;**60**:96-100. DOI: 10.1016/j.phytol.2024.01.017

[8] Danladi MMA, Ogbonna CIC, Ogbonna AI, Giles C, Fletcher MT, Akinsanmi OA. Fungal composition, quantification of mycotoxins, and enzyme activity in processed *Solanum tuberosum Linn* (potato) products stored at different relative humidity. Journal of Applied Microbiology. 2023;**134**(12). DOI: 10.1093/jambio/lxad266

[9] Chen L, Zhang X, Zhang M, Zhu Y, Zhuo R. Removal of heavy-metal pollutants by white rot fungi: Mechanisms, achievements, and perspectives. Journal of Cleaner Production. 2022;**354**:131681. DOI: 10.1016/j.jclepro.2022.131681

[10] Daâssi D, Hajaji AN, Alssulime LJH, Alkhatib SN, Hamouda RA. Fungal consortia mediated bio-treatment of organic matter and metals uptake from sewage water: Maize agro-physiological assessment. Catalysts. 2024d;**14**(4):257. DOI: 10.3390/catal14040257

[11] Ghariani B, Hadrich B, Louati I, Mtibaà R, Daâssi D, Rodriguez-Couto S, et al. Porous heat-treated fungal biomass: Preparation, characterization and application for removal of textile dyes from aqueous solutions. Journal of Porous Materials. 2019b;**26**(5):1475-1488. DOI: 10.1007/s10934-019-00746-6

[12] Swathy K, Vivekanandhan P, Yuvaraj A, Sarayut P, Kim JS, Krutmuang P. Biodegradation of pesticide in agricultural soil employing entomopathogenic fungi: Current state of the art and future perspectives. Heliyon. 2023;**10**(1):23406. DOI: 10.1016/j.heliyon.2023.e23406

[13] Akrout I, Staita K, Zouari-Mechichi H, Ghariani B, Khmaissa M, Navarro D, et al. Valorizing fungal diversity for the degradation of fluoroquinolones. Heliyon. 2024b;**10**(10):e30611. DOI: 10.1016/j.heliyon.2024.e30611

[14] Dinakarkumar Y, Ramakrishnan G, Gujjula KR, Vasu V, Balamurugan P, Murali G. Fungal bioremediation: An overview of the mechanisms, applications and future perspectives. Environmental Chemistry and Ecotoxicology. 2024b;**6**:293-302. DOI: 10.1016/j.enceco.2024.07.002

[15] Navina BK, Velmurugan NK, Kumar PS, Rangasamy G, Palanivelu J, Thamarai P, et al. Fungal bioremediation approaches for the removal of toxic pollutants: Mechanistic understanding for biorefinery applications. Chemosphere. 2024;**350**:141123. DOI: 10.1016/j.chemosphere.2024.141123

[16] Vaksmaa A, Guerrero-Cruz S, Ghosh P, Zeghal E, Hernando-Morales V, Niemann H. Role of fungi in bioremediation of emerging pollutants. Frontiers in Marine Science. 2023;**10**. DOI: 10.3389/fmars.2023.1070905

[17] Gentile L, Floudas D, Olsson U, Persson P, Tunlid A. Fungal decomposition and transformation of molecular and colloidal fractions of dissolved organic matter extracted from boreal forest soil. Soil Biology and Biochemistry. 2024;**195**:109473. DOI: 10.1016/j.soilbio.2024.109473

[18] Mohamed HI, Aal MHA, El-Mahdy OM. Fungal Metabolites and their Role in Remediation of Environmental Pollution. Elsevier eBooks; 2023. pp. 283-315. DOI: 10.1016/b978-0-323-95241-5.00020-4

[19] El-Gendi H, Saleh AK, Badierah R, Redwan EM, El-Maradny YA, El-Fakharany EM. A comprehensive insight into fungal enzymes: Structure, classification, and their role in mankind's challenges. Journal of Fungi. 2021;**8**(1):23. DOI: 10.3390/jof8010023

[20] Dashora K, Gattupalli M, Tripathi GD, Javed Z, Singh S, Tuohy M, et al. Fungal assisted valorisation of polymeric lignin: Mechanism, enzymes and perspectives. Catalysts. 2023;**13**(1):149. DOI: 10.3390/catal13010149

[21] Sun R, Cao Z, Wen X, Ma L, Zhou Y, Li J, et al. Quinone redox cycling drives lignocellulose depolymerization and degradation in composting environments based on metagenomics analysis. The Science of the Total Environment. 2022;**856**:159009. DOI: 10.1016/j.scitotenv.2022.159009

[22] Lin S, Wei J, Yang B, Zhang M, Zhuo R. Bioremediation of organic pollutants by white rot fungal cytochrome P450: The role and mechanism of CYP450 in biodegradation. Chemosphere. 2022;**301**:134776. DOI: 10.1016/j.chemosphere.2022.134776

[23] Daâssi D, Almaghrabi FQ. Petroleum-degrading fungal isolates for the treatment of soil microcosms. Microorganisms. 2023d;**11**(5):1351. DOI: 10.3390/microorganisms11051351

[24] Abdel-Shafy HI, Mansour MS. A review on polycyclic aromatic hydrocarbons: Source, environmental impact, effect on human health and remediation. Egyptian Journal of Petroleum. 2015;**25**(1):107-123. DOI: 10.1016/j.ejpe.2015.03.011

[25] Daâssi D, Almaghribi FQ. Petroleum-contaminated soil: Environmental occurrence and remediation strategies. 3 Biotech. 2022;**12**(6):139. DOI: 10.1007/s13205-022-03198-z

[26] Njoku K, Yussuf A, Akinola M, Adesuyi A, Jolaoso A, Adedokun A. Mycoremediation of petroleum hydrocarbon polluted soil by *Pleurotus pulmonarius*. Ethiopian Journal of Environmental Studies and Management. 2017;**9**(1):865. DOI: 10.4314/ejesm.v9i1.6s

[27] Sharma KR, Giri R, Sharma RK. Efficient bioremediation of metal containing industrial wastewater using white rot fungi. International journal of Environmental Science and Technology. 2022;**20**(1):943-950. DOI: 10.1007/s13762-022-03914-5

[28] Glazunova OA, Moiseenko KV, Fedorova TV. Xenobiotic removal by trametes hirsuta LE-BIN 072 activated carbon-based mycelial pellets: Remazol brilliant blue R case study. Water. 2023;**16**(1):133. DOI: 10.3390/w16010133

[29] Zou JJ, Dai C, Hu J, Tong WK, Gao M, Zhang Y, et al. A novel mycelial pellet applied to remove polycyclic aromatic hydrocarbons: High adsorption performance and its mechanisms. The Science of the Total Environment. 2024;**922**:171201. DOI: 10.1016/j.scitotenv.2024.171201

[30] Mustafa E, Gharieb MM, Mohamed NY. Mycoremediation of industrial wastewater using Aspergillus Niger (EM1) and Fusarium proliferatum (EN1) mats and pellets biomasses. Delta Journal of Science. 2024;**0**(0):0. DOI: 10.21608/djs.2024.306752.1176

[31] Dey P, Malik A, Singh DK, Haange S, Von Bergen M, Jehmlich N. Unveiling fungal strategies: Mycoremediation in multi-metal pesticide environment using proteomics. Scientific Reports. 2024;**14**(1). DOI: 10.1038/s41598-024-74517-y

[32] Paul C, Pal N, Maitra M, Das N. Laccase-assisted bioremediation of pesticides: Scope and challenges. Mini-Reviews in Organic Chemistry. 2022;**21**(6):633-654. DOI: 10.2174/1570193x20666221117161033

[33] Ayed AB, Akrout I, Staita K, Albert Q, Greff S, Simmler C, et al. Genome sequencing of Porostereum spadiceum to study the degradation of levofloxacin. Ecotoxicology and Environmental Safety. 2024;**270**:115808. DOI: 10.1016/j.ecoenv.2023.115808

[34] Ahmed MAEE, Abbas HS, Kotakonda M. Fungal diseases caused by serious contamination of pharmaceuticals and medical devices, and rapid fungal detection using nano-diagnostic tools: A critical review. Current Microbiology. 2023;**81**(1). DOI: 10.1007/s00284-023-03506-7

[35] Arun KB, Madhavan A, Tarafdar A, Sirohi R, AN A, Kuriakose LL, et al. Filamentous fungi for pharmaceutical compounds degradation in the environment: A sustainable approach. Environmental Technology and Innovation. 2023;**31**:103182. DOI: 10.1016/j.eti.2023.103182

[36] Naveen KV, Saravanakumar K, Zhang X, Sathiyaseelan A, Wang M. Impact of environmental phthalate on human health and their bioremediation strategies using fungal cell factory - A review. Environmental Research. 2022;**214**:113781. DOI: 10.1016/j.envres.2022.113781

[37] Saravanan A, Thamarai P, Deivayanai V, Karishma S, Shaji A, Yaashikaa P. Current strategies on bioremediation of personal care products and detergents: Sustainability and life cycle assessment. Chemosphere. 2024;**354**:141698. DOI: 10.1016/j.chemosphere.2024.141698

[38] Shourie A, Vijayalakshmi U. Fungal diversity and its role in mycoremediation.

Geomicrobiology Journal. 2022;**39**(3-5):426-444. DOI: 10.1080/01490451.2022.2032883

[39] Boujelben R, Ellouze M, Tóran MJ, Blánquez P, Sayadi S. Mycoremediation of Tunisian tannery wastewater under non-sterile conditions using Trametes versicolor: Live and dead biomasses. Biomass Conversion and Biorefinery. 2022;**14**(1):299-312. DOI: 10.1007/s13399-022-02328-0

[40] Jasim RAF. Medical, pharmaceutical, and biomedical applications of chitosan. Medical Journal of Babylon. 2021;**18**(4):291-294. DOI: 10.4103/mjbl.mjbl_45_21

[41] Jamir I, Ezung LY, Merry L, Tikendra L, Devi RS, Nongdam P. Heavy metals clean up: The application of fungi for biosorption. Geomicrobiology Journal. 2024;**41**(3):201-212. DOI: 10.1080/01490451.2024.2307899

[42] Li Q, Zhang M, Wei B, Lan W, Wang Q, Chen C, et al. Fungal biomineralization of toxic metals accelerates organic pollutant removal. Current Biology. 2024;**34**(10):2077-2084. DOI: 10.1016/j.cub.2024.04.005

[43] Bibbins-Martínez M, Juárez-Hernández J, López-Domínguez J, Nava-Galicia S, Martínez-Tozcano L, Juárez-Atonal R, et al. Potential application of fungal biosorption and/or bioaccumulation for the bioremediation of wastewater contamination: A review. Journal of Environmental Biology. 2023b;**44**(2):135-145. DOI: 10.22438/jeb/44/2/mrn-5093

[44] Malviya A, Jaspal D. Biosorption Application in Water Reuse and Recovery. Elsevier eBooks; 2023. pp. 93-105. DOI: 10.1016/b978-0-323-99344-9.00007-4

[45] Dhaka R, Kumar S, Kataria N, Kee PE, Ayyamperumal R, Ethiraj B, et al. Biosorption of lead (II) ions by lead-tolerant fungal biomass isolated from electroplating industry effluent. International Journal of Environmental Science and Technology. 2024. DOI: 10.1007/s13762-024-05796-1

[46] Veglio F, Beolchini F. Removal of metals by biosorption: A review. Hydrometallurgy. 1997;**44**(3):301-316. DOI: 10.1016/s0304-386x(96)00059-x

[47] Singh PC, Srivastava S, Shukla D, Bist V, Tripathi P, Anand V, et al. Mycoremediation mechanisms for heavy metal resistance/tolerance in plants. In: Fungal biology. 2018. pp. 351-381. DOI: 10.1007/978-3-319-77386-5_14

[48] Magalhães FF, Pereira AF, Cristóvão RO, Barros RAM, Faria JL, Silva CG, et al. Recent developments and challenges in the application of fungal Laccase for the biodegradation of textile dye pollutants. Mini-Reviews in Organic Chemistry. 2022;**21**(6):609-632. DOI: 10.2174/1570193x20666221104140632

[49] Marco-Urrea E, García-Romera I, Aranda E. Potential of non-ligninolytic fungi in bioremediation of chlorinated and polycyclic aromatic hydrocarbons. New Biotechnology. 2015;**32**(6):620-628. DOI: 10.1016/j.nbt.2015.01.005

[50] Pundir A, Thakur MS, Prakash S, Kumari N, Sharma N, Parameswari E, et al. Fungi as versatile biocatalytic tool for treatment of textile wastewater effluents. Environmental Sciences Europe. 2024;**36**(1). DOI: 10.1186/s12302-024-01007-3

[51] Priyanka N, Dwivedi SK. Fungi mediated detoxification of heavy metals: Insights on mechanisms, influencing factors and recent developments.

Journal of Water Process Engineering. 2023;**53**:103800. DOI: 10.1016/j.jwpe.2023.103800

[52] Khatoon H, Rai J, Jillani A. Role of Fungi in Bioremediation of Contaminated Soil. Elsevier eBooks; 2020. pp. 121-156. DOI: 10.1016/b978-0-12-821925-6.00007-1

[53] Mahmud T, Sabo IA, Lambu ZN, Danlami D, Shehu AA. Hydrocarbon degradation potentials of fungi: A review. Journal of Environmental Bioremediation and Toxicology. 2022;**5**(1):50-56. DOI: 10.54987/jebat.v5i1.681

[54] Tran VV, Park D, Lee Y. Indoor air pollution, related human diseases, and recent trends in the control and improvement of indoor air quality. International Journal of Environmental Research and Public Health. 2020;**17**(8):2927. DOI: 10.3390/ijerph17082927

[55] Daâssi D, Nasraoui-Hajaji A, Bawasir S, Frikha F, Mechichi T. Biodegradation of C20 carbon clusters from diesel fuel by *Coriolopsis gallica*: Optimization, metabolic pathway, phytotoxicity. 3 Biotech. 2021b;**11**(5). DOI: 10.1007/s13205-021-02769-w

[56] Azin E, Moghimi H, Heidarytabar R. Petroleum degradation, biosurfactant and Laccase production by Fusarium neocosmosporiellum RH-10: A microcosm study. Soil and Sediment Contamination an International Journal. 2018;**27**(4):329-342. DOI: 10.1080/15320383.2018.1473334

[57] Nrior RR, Onwuka NF. Bioremediation of crude oil contaminated marshland muddy soil by bioaugmentation approach using *Candida tropicalis* and *Penicillium chrysogenum*. IOSR Journal of Environmental Science, Toxicology and Food Technology. 2017;**11**:2319-2399

[58] Rathankumar AK, Saikia K, Ramachandran K, Batista RA, Cabana H, Vaidyanathan VK. Effect of soil organic matter (SOM) on the degradation of polycyclic aromatic hydrocarbons using Pleurotus dryinus IBB 903-a microcosm study. Journal of Environmental Management. 2020;**260**:110153. DOI: 10.1016/j.jenvman.2020.110153

[59] Yanto DHY, Hidayat A, Tachibana S. Periodical biostimulation with nutrient addition and bioaugmentation using mixed fungal cultures to maintain enzymatic oxidation during extended bioremediation of oily soil microcosms. International Biodeterioration and Biodegradation. 2016;**116**:112-123. DOI: 10.1016/j.ibiod.2016.10.023

[60] Košnář Z, Částková T, Wiesnerová L, Praus L, Jablonský I, Koudela M, et al. Comparing the removal of polycyclic aromatic hydrocarbons in soil after different bioremediation approaches in relationto the extracellular enzyme activities. Journal of Environmental Sciences. 2018;**76**:249-258. DOI: 10.1016/j.jes.2018.05.007

[61] Godambe T, Fulekar M. Bioremediation of petrochemical hydrocarbons (BTEX) – Review. The Journal of Environmental Science Pollution Research. 2017;**3**(3):189-199

[62] Bhardwaj G. Biosurfactants from fungi: A review. Journal of Petroleum and Environmental Biotechnology. 2013;**04**(06). DOI: 10.4172/2157-7463.1000160

[63] Khan R, Patel V, Khan Z. Bioremediation of Dyes from Textile and Dye Manufacturing Industry Effluent. Elsevier eBooks; 2019. pp. 107-125. DOI: 10.1016/b978-0-12-818095-2.00005-9

[64] Madani VDM. The study of heterotrophic and crude oil-utilizing soil fungi in crude oil contaminated regions. Journal of Bioremediation and Biodegradation. 2015;**06**(02). DOI: 10.4172/2155-6199.1000270

[65] Da Silva Júnior FMR, Volcão LM, Hoscha LC, Pereira SV. Growth of the fungus *Chaetomium aureum* in the presence of lead: Implications in bioremediation. Environmental Earth Sciences. 2018;**77**(7). DOI: 10.1007/s12665-018-7447-x

[66] Chan-Cupul W, Heredia-Abarca G, Rodríguez-Vázquez R. Atrazine degradation by fungal co-culture enzyme extracts under different soil conditions. Journal of Environmental Science and Health Part B. 2016b;**51**(5):298-308. DOI: 10.1080/03601234.2015.1128742

[67] Ray MK, Panda J, Panda BP, Mohanta TK, Mohanta YK. Mycoremediation of heavy metals and/or metalloids in soil. In: Sarma H, Joshi S, editors. Land Remediation and Management: Bioengineering Strategies. Singapore: Springer; 2023. DOI: 10.1007/978-981-99-4221-3_8

[68] Sepehr MN, Zarrabi M, Amrane A, Samarghandi MR. Removal of Cr (III) from model solutions and a real effluent by *Phanerochaete chrysosporium* isolated living microorganism: Equilibrium and kinetics. Desalination and Water Treatment. 2013;**51**(28-30):5627-5637. DOI: 10.1080/19443994.2013.769698

[69] Castaño-Ortiz J, Gil-Solsona R, Ospina-Álvarez N, Alcaraz-Hernández J, Farré M, León V, et al. Fate of pharmaceuticals in the Ebro River Delta region: The combined evaluation of water, sediment, plastic litter, and biomonitoring. The Science of the Total Environment. 2023;**906**:167467. DOI: 10.1016/j.scitotenv.2023.167467

[70] Esterhuizen-Londt M, Schwartz K, Pflugmacher S. Using aquatic fungi for pharmaceutical bioremediation: Uptake of acetaminophen by Mucor hiemalis does not result in an enzymatic oxidative stress response. Fungal Biology. 2016;**120**(10):1249-1257. DOI: 10.1016/j.funbio.2016.07.009

[71] Baran W, Adamek E, Włodarczyk A, Lazur J, Opoka W, Muszyńska B. The remediation of sulfonamides from the environment by *Pleurotus eryngii* mycelium. Efficiency, products and mechanisms of mycodegradation. Chemosphere. 2020;**262**:128026. DOI: 10.1016/j.chemosphere.2020.128026

[72] Nosek K, Zhao D. Transformation products of diclofenac: Formation, occurrence, and toxicity implication in the aquatic environment. Water Research. 2024;**266**:122388. DOI: 10.1016/j.watres.2024.122388

[73] Copete-Pertuz LS, Plácido J, Serna-Galvis EA, Torres-Palma RA, Mora A. Elimination of Isoxazolyl-Penicillins antibiotics in waters by the ligninolytic native Colombian strain *Leptosphaerulina sp*. considerations on biodegradation process and antimicrobial activity removal. The Science of the Total Environment. 2018;**630**:1195-1204. DOI: 10.1016/j.scitotenv.2018.02.244

[74] Del Álamo AC, Pariente M, Martínez F, Molina R. Trametes versicolor immobilized on rotating biological contactors as alternative biological treatment for the removal of emerging concern micropollutants. Water Research. 2019;**170**:115313. DOI: 10.1016/j.watres.2019.115313

[75] Raudabaugh DB, Miller AN, Gunsch CK. Evaluation of pesticides on detritus-inhabiting and root-associated fungi in aquatic habitats and potential

implications. Diversity. 2024;**16**(5):255. DOI: 10.3390/d16050255

[76] Matúš P, Littera P, Farkas B, Urík M. Review on performance of *Aspergillus* and *Penicillium* species in biodegradation of organochlorine and organophosphorus pesticides. Microorganisms. 2023;**11**(6):1485. DOI: 10.3390/microorganisms11061485

[77] Gajendiran A, Abraham J. Biomineralisation of fipronil and its major metabolite, fipronil sulfone, by Aspergillus glaucus strain AJAG1 with enzymes studies and bioformulation. 3 Biotech. 2017;**7**(3). DOI: 10.1007/s13205-017-0820-8

[78] Maheshwari A, Srivastava S, Srivastava S. Fungi as a Tool for Decontaminating the Range of Soil Contaminants. Elsevier eBooks; 2024. pp. 189-226. DOI: 10.1016/b978-0-443-19049-0.00004-9

[79] Spinelli V, Ceci A, Bosco CD, Gentili A, Persiani AM. Glyphosate-eating fungi: Study on fungal saprotrophic strains' ability to tolerate and utilise glyphosate as a nutritional source and on the ability of *Purpureocillium lilacinum* to degrade it. Microorganisms. 2021;**9**(11):2179. DOI: 10.3390/microorganisms9112179

[80] Saroj S, Kumar K, Pareek N, Prasad R, Singh R. Biodegradation of azo dyes acid red 183, direct blue 15 and direct red 75 by the isolate *Penicillium oxalicum* SAR-3. Chemosphere. 2014;**107**:240-248. DOI: 10.1016/j.chemosphere.2013.12.049

[81] Mohamed ZA, Hashem M, Alamri S, Campos A, Vasconcelos V. Fungal biodegradation and removal of cyanobacteria and microcystins: Potential applications and research needs. Environmental Science and Pollution Research. 2021;**28**(28):37041-37050. DOI: 10.1007/s11356-021-14623-w

[82] Ngieng NS, Zulkharnain A, Roslan HA, Husaini A. Decolourisation of synthetic dyes by endophytic fungal flora isolated from senduduk plant (*Melastoma malabathricum*). ISRN Biotechnology. 2013;**2013**:1-7. DOI: 10.5402/2013/260730

[83] Sen S, Raut S, Raut S. Mycoremediation of anthraquinone dyes from textile industries: a mini-review. Biotechnologia. 2023;**104**(1):85-91. DOI: 10.5114/bta.2023.125088

[84] Negi BB, Das C. Mycoremediation of wastewater, challenges, and current status: A review. Bioresource Technology Reports. 2023;**22**:101409. DOI: 10.1016/j.biteb.2023.101409

[85] Przystaś W, Zabłocka-Godlewska E, Grabińska-Sota E. Efficiency of decolorization of different dyes using fungal biomass immobilized on different solid supports. Brazilian Journal of Microbiology. 2017;**49**(2):285-295. DOI: 10.1016/j.bjm.2017.06.010

[86] Dasgupta D, Barman S, Sarkar J, Mridha D, Labrousse P, Roychowdhury T, et al. Mycoremediation of different wastewater toxicants and its prospects in developing value-added products: A review. Journal of Water Process Engineering. 2024;**58**:104747. DOI: 10.1016/j.jwpe.2023.104747

[87] Harms H, Schlosser D, Wick LY. Untapped potential: Exploiting fungi in bioremediation of hazardous chemicals. Nature Reviews Microbiology. 2011;**9**(3):177-192. DOI: 10.1038/nrmicro2519

[88] Fan B, Zhao Y, Mo G, Ma W, Wu J. Co-remediation of DDT-contaminated soil using white rot fungi and laccase

extract from white rot fungi. Journal of Soils and Sediments. 2013;**13**(7):1232-1245. DOI: 10.1007/s11368-013-0705-3

[89] Latif W, Ciniglia C, Iovinella M, Shafiq M, Papa S. Role of white rot fungi in industrial wastewater treatment: A review. Applied Sciences. 2023;**13**(14):8318. DOI: 10.3390/app13148318

[90] Velmurugan P, Shim J, You Y, Choi S, Kamala-Kannan S, Lee K, et al. Removal of zinc by live, dead, and dried biomass of Fusarium spp. isolated from the abandoned-metal mine in South Korea and its perspective of producing nanocrystals. Journal of Hazardous Materials. 2010;**182**(1-3):317-324. DOI: 10.1016/j.jhazmat.2010.06.032

[91] Zuo Y, Chen G, Zeng G, Li Z, Yan M, Chen A, et al. Transport, fate, and stimulating impact of silver nanoparticles on the removal of Cd(II) by *Phanerochaete chrysosporium* in aqueous solutions. Journal of Hazardous Materials. 2014;**285**:236-244. DOI: 10.1016/j.jhazmat.2014.12.003

[92] Jakubiak M, Giska I, Asztemborska M, Bystrzejewska-Piotrowska G. Bioaccumulation and biosorption of inorganic nanoparticles: Factors affecting the efficiency of nanoparticle mycoextraction by liquid-grown mycelia of Pleurotus eryngii and Trametes versicolor. Mycological Progress. 2013;**13**(3):525-532. DOI: 10.1007/s11557-013-0933-3

[93] Salvadori MR, Ando RA, Nascimento CAO, Corrêa B. Extra and intracellular synthesis of nickel oxide nanoparticles mediated by dead fungal biomass. PLoS One. 2015;**10**(6):e0129799. DOI: 10.1371/journal.pone.0129799

[94] Hussain K, Haris M, Qamar H, Hussain T, Ahmad G, Ansari MS, et al. Bioremediation of waste gases and polluted soils. Microorganisms for Sustainability. 2021:111-137. DOI: 10.1007/978-981-15-7455-9_5

[95] Hashmi MZ, Kumar V, Varma A. Xenobiotics in the soil environment. Soil Biology. 2017. DOI: 10.1007/978-3-319-47744-2

[96] Rhodes CJ. Mycoremediation (bioremediation with fungi) – Growing mushrooms to clean the earth. Chemical Speciation and Bioavailability. 2014;**26**(3):196-198. DOI: 10.3184/095422914x14047407349335

Chapter 6

The Role of Bioremediation in Achieving Environmental Sustainability

Wafaa M. Abd El-Rahim and Hassan Moawad

Abstract

Research on biological solutions for sustainable environmental health has grown significantly over the years, highlighting the urgent need for innovative and effective remediation strategies. Bioremediation, an ecologically significant practice, uses biological agents to address despoiled habitats by harnessing the metabolic potential of microorganisms or cells to degrade, remove, or dispose of contaminants from soil, water, or air. Bioremediation pathways are part of general life processes, like energy production, and do not result in the breaking down of one pollutant into a harmful substance. A broad spectrum of pollutants can be bioremediated through a specific selection of biological agents such as bacteria, microalgae, aquatic pulmonate, fungi, and plants. The ecological health of any ecosystem and human health are emphasized, as chemical spills into environmental components can influence sustainable management hydraulics and ecosystem results due to the intricate food chain. Bioremediation stands as a potential development and low-cost methodology for removing pollutants heavily from affected parts, and several cases of microorganisms and their enzymes and processes are used in the removal of industrially produced hazardous substances. The following sections are dedicated to a detailed understanding of microorganisms and exploration of the potential of biodegradation processes useful for the removal of environmental pollutants, as well as the reduction in human health hazards. Bioremediation stands as one of the potential developments and low-cost methodologies for removing pollutants heavily from the affected parts. Furthermore, several cases of microorganisms and their enzymes and processes involved are used in the removal of industrially produced hazardous substances.

Keywords: bioremediation, living organisms, sustainable, eco-friendly, environmental challenges

1. Introduction

Bioremediation is an innovative and effective process that employs natural biological systems to clean up and restore contaminated environments effectively. This method stands out as a low-cost and environmentally friendly solution that holds significant potential for promoting sustainable development across various regions.

The remarkable ability to clean up and rehabilitate contaminated environments helps restore natural ecosystems, effectively transforming an area that was once considered a liability into a valuable asset. Bioremediation represents a progressive technology that harnesses the process of contaminant degradation, utilizing living organisms to convert harmful metal contaminants into less toxic, less soluble, and/or immobile forms that pose minimal risk to the ecosystem. While the bioremediation of inorganic contaminants is still not widely embraced or accepted, some regions have begun to adopt it on a trial and experimental basis, reflecting a growing interest in this viable solution. In general, it is essential to regard bioremediation not merely as a treatment technology but also as a vital journey toward comprehensive ecosystem restoration and alignment with the principles of green chemistry, which emphasize sustainability and environmental stewardship. Through careful implementation and ongoing research, bioremediation can play a crucial role in healing our damaged landscapes and promoting environmental health. Bioremediation makes use of living organisms to remove or neutralize contaminants. Bioremediation technologies use naturally occurring bacteria, fungi, plants, and other organisms to degrade or assimilate toxic substances into non-toxic substances. Fungi have been used to clean up a variety of environmental contaminants from chlorinated solvents to hydrocarbons. Early bioremediation efforts, many of which were quite successful, focused on biostimulation strategies. These can vary considerably, from simple strategies to complex, elaborate ones. While it is not necessary to dig up large amounts of soil for solid-phase treatments, it is important to keep in mind the geographical and geological context and to conduct appropriate feasibility studies. Although people have used bioremediation techniques at least since Roman times, the modern bioremediation industry is a product of the last 50 years as well as new concepts in environmental chemistry and toxicology. This growth is in part a response to the vast increase in the number of contaminated sites caused by industrial growth. The import stems are increasing faster than clean systems [1, 2].

2. Definition and principles

The term bioremediation is also often used as a synonym for the process of using biological organisms to detoxify polluted environments. There are also various definitions available for bioremediation. However, the following description is a practical and working definition: bioremediation is the process of using biological processes to harness the capacity of biological systems to degrade, sequester, or immobilize environmental contaminants to protect human health, animal health, or whole ecosystems. To perform the process assessment, the main factors that can affect remedial capabilities and toxic levels in a sample are taken into consideration, which include environmental and biological factors. The effectiveness of microbial processes is firmly established through preceding research. When microorganisms break down compounds, they use metabolic pathways to form a variety of intermediate compounds and ultimately generate carbon dioxide, water, inorganic salts, and additional microorganisms. These ultimate advantages of microbial metabolism form the esthetic appeal of bioremediation. Microorganisms also have a greater metabolic capability than higher organisms because they are capable of degrading and/or detoxifying lipophilic organic compounds. Microorganisms can metabolize these compounds through the degradative or catabolic pathways. Though metabolism primarily results in the biotransformation of the original pollutant into another organic form, it is also possible that microorganisms can degrade to inorganic products by mineralization or restoration of the original compound structure. Although it is not certain, and also depends on the level of metabolism, soil and environmental conditions can cause the converted compounds to accumulate and, in turn, increase toxicity [3].

3. Historical development

The emergence of industrial microbiology known as the fermentation industry in which microbial processes are used to obtain chemicals, pharmaceuticals, and fuels in addition to the manufacture of food products, resulted in development of bioprocesses for the treatment of wastewater, industrial emissions, and solid waste from multiple sources including Abattoirs, Dairy, Pulp and paper, Pharmaceuticals, Petro-chemicals, Energy and Textiles have undergone judicious intervention with bacteria, fungi, yeasts, and algae. Although biotechnological processes were developed to control, optimize and mitigate large scale fermentation procedures to break down and detoxify the environment of sugar, amino acid, and alcohol fortified effluents, the organic compounds such as lignin and humic substances were only structurally unaltered and hence multifaceted in toxicity. This led to the development of technology for secondary wastewater treatments based on physico-chemical processes, which employed acclimatized bacteria from polluted environments, giving rise to "activated sludge technology." Facultative aerobic bacteria within activated sludge, aerobic activated sludge for the consumption of external organic substrates that occur within the same cells constituting a "contained microenvironment." This ended the era of sterile bacteriology, starting modern biological methods for controlling bioprocesses and addressing environmental issues [4]. Bioremediation is not a new concept. Historically, this process had been observed among primitive agricultural societies, which tweaked the microbial communities present in the soil to treat pollution. Industrial Revolution marked the beginning of a new era of exploitation of natural resources *via* anthropogenic activities such as mining and the use of petroleum, which resulted in an unprecedented number of pollutants being released into the environment [5]. There are numerous examples that illustrate the destruction caused by anthropogenic pollution. One such example is the contamination of the Minimata Bay in the former industrial city of Minimata in Japan. It was noted that the wastewater discharged from the Chisso chemical factory contained high levels of methylmercury as a byproduct. The levels increased when electrical production due to hydroelectric dams led to Hg^{2+} that was being deposited in the Minimata Bay being converted to CH_3Hg^+ through biotic interaction, and got assimilated by the marine biota which became toxic to humans and the cats that were higher up in the marine trophic levels [6].

4. Types of bioremediation techniques

Bioremediation can be categorized into four primary methodologies. These are *in situ* bioremediation, *ex situ* bioremediation, biostimulation, and bioaugmentation. *In situ* bioremediation is the unobtrusive treatment of a contaminated site;

the contaminants are treated within the site without being removed or drawn to the surface. In certain usages, this term represents both *in situ* and on-site bioremediation. In this instance, biological, chemical, or other treatment systems are set up on-site to treat contaminated soil, water, or air. The general idea is to bring the treatment to the contamination to maximize efficiency and minimize disturbance. Conversely, *ex situ* bioremediation is microbially stimulated in the straightforward treatment of contaminated material that is physically removed from the initial site. It is then treated repeatedly and, once clean, returned to the site en masse. *Ex situ* bioremediation is often used at steeply contaminated sites where on-site cleanup is not obvious [2, 7–9]. Biostimulation aims to encourage extant populations of microorganisms to degrade contaminants more quickly. This works by providing them with all of the necessary nutrients. If the right nutrients are supplied, the native strains of microorganisms can break down organic contaminants (up to their capacity) with biostimulation. Alternatively, bioaugmentation involves the addition of microorganisms to the ecosystem to increase the numbers of organisms able to degrade the contaminants. In the proper circumstances, bioaugmentation can focus the activity of specialized microorganisms to clean up the contamination. Bioremediation can be geared to aquatic or terrestrial systems, and it can differ in the source or mode of introduction of organisms. In land farming, now a relatively common technique, contaminated soil is overturned for periods of time and moisture is checked periodically; this creates an environment where biodegradative bacteria can propagate and metabolize the contaminants. In soil vapor extraction, petroleum hydrocarbons are volatilized by the extraction of volatile interferents into the air stream to facilitate uptake by microorganisms for biodegradation. The removal of aromatics by bacteria, which are reconnected to at least 50% of the oil from the ground, is another form of *ex situ* remediation. The resilience of biodegradable microorganisms to competition from aerosolized microbes generated through steam cleaning or chemical dispersion has been established in a variety of trials. However, *ex situ* cultures did not survive when reintroduced. Bioaugmentation, usually used with composting, is a type of bioremediation in which the concentration of microorganisms in the soil is enhanced to treat contaminated soils. Bioaugmentation, on the other hand, is typically considered to be a relatively unsustainable way to treat environmental legacies because the newly introduced species cannot always survive under the often-harsh environmental conditions. In contrast, in regions with elevated concentrations of polychlorinated biphenyls, *in situ* indigenous forces have achieved greater success [10–12].

5. Microorganisms in bioremediation

The bioremediation field has seen significant numbers of new bacteria discovered that can break down oil and use its aromatic hydrocarbons as an energy source. There is substantial commercial interest in these microorganisms, which can be used to treat soils contaminated with crude oil. Their production can be used to ascertain the environmental fate and half-life of such compounds. Their activity can be used in both the bioremediation of benzene-mediated pollution in soils and in the refining and provision of additional natural resources such as biofuels and recombinant microbial bioremediation. There is also interest in assessing whether methods used to biologically produce fine chemicals and pharmaceutical products can be adapted for bioremediation purposes where there is modular employment of engineered bacterial enzymes. So, for example, DDT was added to culture of *Pseudomonas putida* JMP134;

subsequently, its catabolic processes were used to purify the compound. These organisms can also shrink and discharge antibodies that can detect harmful waste, which could be used as biosensors for on-site rapid discovery of microorganisms that are central to the bioremediation process. They are known to have great potential in the field of biodegradation, which is the degradation of various types of environmental pollution by natural means. According to this definition, biodegradation could be an advantage to clean up the polluted areas. The geographic location of the polluted area will determine the type of microbial community responsible.

Microorganisms can be classified into bacteria, fungi, actinomycetes, and algae groups and are widely distributed throughout the soil or wastewater, exhibiting a unique role in a single enzyme. For example, bacteria play a role in different metabolic activities. These groups have strains that are distinguished by a type of enzyme that will degrade a given pollutant. Another possibility of these microorganisms is that the microbial community responsible has the ability to interact with each other during pollutant degradation; one bacterial strain can supply a given substance for the growth of the related strain of bacteria. There is a diverse assem [13–15] blage of microorganisms capable of biodegradation. Numerous microorganisms, like bacteria, fungi, yeast, and other single-celled eukaryotes, have been identified for rapid growth and degrading various toxins present in the polluted soil or liquid medium. Consequently, a number of well-characterized microorganisms have attracted much attention in laboratories for their greater ability to bio-utilize and provide energy to the associated cells, if any, for biodegradation. These microbes will become fully active when they are provided with a high content of nutrients such as carbon and energy sources. When conditions are suitable, chemicals are converted to the intermediate, cell biomass, and gaseous and energy compounds in accordance with the typical bacterial metabolic pathway. The microbe-mediated biodegradation of the PAHs will occur at the required pH range. Typically, a pH of 5 to 10 is suitable for the growth of various bacteria.

6. Key microbial species

Bioremediation is the ability of microorganisms to degrade, alter, and metabolize pollutants found in environments. Key microbial players used in bioremediation projects include many species of bacteria, fungi, yeast, and, recently, archaeal species. Members of all three domains of life possess the ability to biodegrade various compounds; fungi and bacteria are the most frequently used for the degradation of xenobiotics such as polycyclic hydrocarbons, polychlorinated biphenyls, polychlorinated dibenzo-p-dioxins, ethidium bromide, reactive dyes, and some herbicides and pesticides. Bacterial species are usually employed for the biodegradation of hydrocarbons, heavy metals, organic, and inorganic pollutants, whereas fungi are mainly used to degrade lignins, polycyclic hydrocarbons, cellulose, hemicellulose, and toxic aromatic compounds [15–17]. Bacterial species offer several advantages, such as producing various extracellular enzymes responsible for the bioremediation of a vast range of oligotrophic environments; interacting with available nutrient sources as well as being amenable to culturing under laboratory conditions; being genetically manipulatable with respect to isolation, cloning, and transferring of genes of interest; exhibiting their presence even if targets are found in the groundwater; and conferring protective protein parts for the further generation of recombinant products of bacteria. The first bacterial species was isolated from the soil, while the fungal species live mostly on decaying organic matter and can be found in the soil, in the colon of certain animals, and in association with plants in the form of mycorrhizal species. Archaea are prokaryotic microorganisms closely related to bacteria and are found in the most extreme environments, like the gut of living beings and in geothermal habitats as the most heat-resistant species. Genomic studies of uncultivable microorganisms are necessary to verify their capabilities for their natural application in bioremediation. Bioaugmentation is done after the isolation, culturing, and augmentation of the organism. Bioaugmentation is advantageous when a lake ecosystem is to be cleaned in less time with fewer operational strategies [18–20].

7. Microbial community dynamics

Biodegradation as a microbial utilization of pollutants led to the two cooperatively acting components, substrates and enzymes. Substrates are very specific in their mode of interactions with the biological macromolecules and thus act as a driving force to select an enzyme for a particular reaction. As enzymes are proteins or their macromolecular complexes, their catalytic properties can be modulated in response to structural and conformational modifications to the interaction with pollutants. The structure of the microbial community dynamics can be predicted from the given environmental conditions and functional potential derived from the functional structure. Ecosystem functions are often not determined by the abundance of the individual species, but by species interaction. Interaction between the organisms is influenced by biotic and abiotic factors. In natural environments, the structures of the microbial communities can change actively through secondary metabolites, biofilms, and MDRs or passively through random genetic drift, founder events, bottlenecks, and population sizes due to evolution. Different consortia can have a significant effect on pollutant degradation and therefore, BAS processes. The implication, however, is that environmental conditions may affect the structure of the community and, consequently the biodegrader's structure. The number and rate of metabolism, multiplication, copies of MDRs and toxins, and formation of exoenzymes by the microbes are increased or decreased by the addition of nutrients. Likewise, environmental factors can affect the diversity and dynamics of microbial activities, which is the subject of ecophysiology. Therefore, there is a need to study the dynamics of microbial communities, which are involved in pollutant cold assimilation and its interaction [21–24].

8. Enzymes in bioremediation

Enzymes are the most effective and selective biocatalysts in bioremediation. Enzymes accelerate the breakdown of complex and scientifically recalcitrant pollutants into simpler and less toxic metabolites. By affecting cell membrane permeability, the enzymes improve the cell's ability to degrade organic pollutants and assist in solubilizing hydrophobic contaminants. Enzymes catalyze every conceivable biological hydrolytic, reductive, oxidative, and conjugative reaction. There are different types of enzymes involved in catalysis, including hydrolases, involved in hydrolysis; oxidoreductases, involved in oxidation-reduction activities; isomerases, involved in isomerization; and transferases, involved in transferring chemical units from one molecule to another. Microbial enzymes act on the hydrophobic molecules to form hydrophilic products. When microorganisms reach the boundary of the hydrophobic pollutants in the contaminated zone, an enzyme is secreted, serving as a partial reaction in the process, splitting the contaminated sites into hydrophilic products that can be assimilated into the cell. Then the products can be used as carbon, carrier, and electron donor to transform to CO_2 and H_2O [25–27]. Different types of bacterial strains have their own associated enzymatic activities that assist in carrying out the process of bioremediation efficiently. This is an association between the bacteria and the enzymes, where the bacteria secrete enzymes that are active against a majority of the pollutants present in the particular area. Based on the characteristics of the microorganisms and the enzymes present, several types of bioremediation can be carried out efficiently. Important enzymes involved in the degradation of complex hydrocarbons to simple hydrocarbons include dehalogenases, alcohol dehydrogenases, catechol dioxygenase, catechol 1,2-dioxygenase, aromatic ring cleavage, protocatechuate meta-cleavage, salicylate hydroxylase, 1-hydroxyl-2-naphthoate dioxygenase, and others. Crop root exudates can act as a natural magnetic field for the microbes, while the hyperaccumulators act as a magnet for the remediators. Most of the paddy crops have a rhizosphere that sludge sticks to the surface of the roots, whereas the area above the roots is waterlogged. Many of the steps in the bioremediation process are catalyzed by enzymes. The efficiency and stability of enzymes used in practical remediation applications have always been a problem, but we can comfort ourselves that many improvements can be made using a variety of novel approaches, including protein engineering. Many natural biological processes are governed by the constraints of specific enzymes produced and degraded or secreted by the particular microbial species that live in a particular environmental condition. The specificity of these enzymes and the conditions requiring their production contribute to the overall selectivity and strategies microbial species use to coexist and produce functional proteins [28–30].

9. Enzymatic processes

The potential of microorganisms to degrade pollutants results from their association in the global environment over millions of years. The enzymatic processes responsible for the biotransformation and mineralization of organic pollutants depend on the type of compounds as well as the enzymes involved in their transformation to potentially less and non-toxic compounds. The chemical nature of organic pollutants and their susceptibility to microbial decomposition may be dependent on the structure, complexity, and functional groups available for enzymatic transformation. The main categories of enzymatically catalyzed reactions encountered in the bioremediation of organic compounds based on nitrogen include oxidation and oxygenation; reduction; hydrolysis/lyase oxidation; methanogenesis; dehydrogenation/hydroxylation/oxidation-reduction; and transferase, mutase, and lyase reactions [31–35]. Chemical pollutants can be classified as xenobiotics, non-natural compounds lacking in environmental samples resulting from industrial effluents. The majority of these compounds are lipophilic and often toxic due to the unevaluated compounds in the industrial manufacturing process and potential derivatives from convenient precursors. The solution to convert industrial wastes into reliable and less toxic derivatives lies in the form of bioremediation. Enzymatic degradative processes are very effective in the degradation of virtually all environmental pollutants present in the environment. Moreover, the results from the use of appropriate enzymes have excellent value in the minimization of environmental pollution and detoxification. However, the use of endogenous and exogenous enzymes is scarce in the literature and needs more exploration [36–38].

10. Enzyme production and optimization

Enzyme sources enzymes can be obtained from microbial, plant, and synthetic sources. Microorganisms are the primary sources of enzymes used in bioremediation. In particular, bacteria, fungi, and yeast are the main enzyme-producing microorganisms. In biotechnological applications, enzymes can also be produced from plant or animal sources as well as from genetically modified microorganisms. Compared to plant-derived enzymes, the main advantage of microbial enzymes is the potential for faster and larger scale cultivation using fermentation and bioprocessing techniques. For these reasons, most research on enzyme production has focused on microbial sources. Enzymes produced from genetic sources can be improved by genetic and biotechnological modifications. Although enzymes are usually produced by chromatographic and crystallization purification, immobilization techniques can be employed for enzyme separation [13, 38–40]. Enzyme production optimization of enzyme production can improve enzyme yield, activity, and economics in bioremediation. Efforts have been made to improve enzyme production by using advanced molecular tools, including molecular biology and genetic and metabolic engineering. The optimization of enzyme production typically involves optimizing enzyme activity, expression systems, and culture conditions as well as scaling up the production processes. Optimization of these parameters can maximize the expression of high-quality enzymes during the production process. Producing industrial-scale quantities of high-quality enzymes requires a larger quantity of low-cost raw materials and, in turn, results in lower cost of production and increased competition with other treatments. Some fermentation studies have shown that, after production optimization, the yield of crude enzyme increased and provided more optimal results than the amount of enzyme expressed in normal conditions. One of the most successful changes in the development of fermentation strategies is to increase the capacity of the microbial system to express crude protein through the semi-purification of the active enzyme mixture [41–43].

11. Applications of bioremediation

Bioremediation has the potential to remediate a wide variety of environmental pollution in various ecosystems, including soils, water, and air. For example, bioremediation can be used to restore soil or water contaminated with mining or military activities, or to restore or rehabilitate oil spills in oceans or seas. For this topic, bioremediation serves all ecosystems. Some successful case studies of bioremediation applications in different ecosystems are described next. In the most unconventional environments, like the Arctic ice, bioremediation techniques are not commonly applied. However, a bioremediation process was able to counteract hydrocarbon pollution in North-West Spitsbergen. In the desert, a consortium of bacteria able to degrade PAHs and phenols has been selected with favorable applications in bioremediation. More conventional applications of bioremediation are described, for instance, for cleaning oil-polluted water or groundwater, using, in some cases, immobilized bacteria or simple equipment to help the process. The same applies to the oxic treatment of groundwater using bacteria or the rehabilitation of mercury-polluted sites with one strain of bacteria. As in the case studies, bioremediation is mainly applied in the restoration of contaminated sites since it contributes to the ecological resiliency of these sites and, in addition, improves the local inhabitants' public health. This approach not only facilitates the removal of toxic substances but also mitigates the adverse effects on the surrounding ecosystem. Additionally, the use of specific microbial strains can enhance the efficiency of degradation processes, making bioremediation a viable solution in various environmental contexts. Recent advancements in genetic engineering further allow for the development of engineered bacteria that can target and break down complex pollutants more effectively highlight the importance of microbial processes in degrading pollutants, such as hydrocarbons and heavy metals, which are crucial for restoring contaminated environments [44]. Bioremediation techniques are highly versatile since they can be applied for the removal of a large variety of materials, including organic and inorganic compounds, such as heavy metals, emerging pollutants like perfluorooctanoic acids, and nanoparticles, among others. Furthermore, it can be applied to a wide variety of environmental conditions, from the cleaning of contaminated aquifers to the restoration of soils or the polluted atmosphere in urban areas. Indeed, bioremediation can take several forms depending on the environmental conditions and the matrix where the pollutants are found. When the bioremediation technique is applied to remove contaminants from a liquid medium, whether it is water or soil extracted with the appropriate solvents, the technique is called biosorption and bioaccumulation, respectively. An example of a successful application of these techniques is the removal of heavy metals from industrial wastewater. In the case of sites contaminated with organic pollutants, the bioremediation technique can be applied to the rehabilitation of the soil, and it is used to stimulate the degradation of the different contaminants by the endogenous microbial population. This group is represented by several other possibilities, such as *in situ*, on site, surface, bioaugmentation, biostimulation, constructed wetlands, or hotspots, among others. Additionally, the bioremediation technique can be applicable at all stages of the pollution; for example, when a decontamination procedure is required, it can be applied at the time the contamination is produced or after. The simultaneous application is known as phytoremediative bacterio-phytostabilization, while the residual application is known as a biostabilization mechanism [45–47].

12. Soil remediation

Bioremediation has enormous potential for soil remediation. Heavy metals, pesticides, and petroleum products represent the main contaminants found in soils. Several strategies can allow contaminant degradation by both autochthonous and allochthonous organisms. Among them, phytoremediation, when assisted by plant-associated microorganisms, and microbially-assisted remediation have proven to be effective in decontaminating polluted areas. Crops are grown in soils, with preference given to those that produce biomass with high energy density. Remediated areas are often left degraded and polluted, and the interest in restoring soil health has recently increased. Bioremediation often takes place in grasslands and forests. Soil fatty acids have been revised, and different applications have been presented in the emerging world of soil. Many examples of successful soil remediation with both bacterial and fungal amendments or *in situ* stimulation have been demonstrated. The use of local microorganisms is considered a valid alternative to the use of allochthonous bacteria to exert biostimulation and bioaugmentation. Consistently with this, bioremediation of contaminated soil is depicting bioremediation strategies based on local microflora exploiting several *in situ* technologies. After each bioremediation application, both the residual pollutant content and any potential negative impacts require the use of sophisticated analytical and assessment strategies to determine and evaluate the effectiveness of monitored technologies [48–51].

13. Water treatment

Bioremediation strategies have been of great interest for the treatment of waters contaminated by a wide array of biodegradable organic and inorganic compounds. In such cases, the main processes that occur are degradation of the pollutants through chemical and biological reactions. Microorganisms, particularly bacteria, are the key players in such remediation projects. In subsurface environments with anoxic conditions, microorganisms degrade pollutants in the absence of oxygen. This process, known as anaerobic bioremediation, can also be stimulated using a number of different approaches [52].

The concentration of pollutants in contaminated water may be quite low, such as nutrients, making bioremediation by microorganisms problematic. In such a case, a strategy called biostimulation is typically used to increase the numbers of indigenous microorganisms. Alternatively, in some cases, the contamination may be so severe that the indigenous microorganisms are unable to degrade the pollutants under normal conditions. In such cases, bioaugmentation may also be utilized. In the case of solid or hazardous waste landfills, bioaugmentation may involve the artificial introduction of either chosen or engineered microorganisms that could potentially degrade the contaminants. Contaminants in water are typically of two types: either organic compounds like chlorinated organic solvents, coal derivatives, and chlorophenols, or inorganic compounds such as heavy metals, nitrogen, phosphorus, dieldrin, and other non-specific organic compounds. Based on the biodegradability of the pollutants, different strategies have been developed for remediating contaminated water [53–56]. The large surface of rivers, lakes, estuarine systems, and seas is in sustained contact with soil or sediments at various degrees. Microorganisms are present in large numbers together with invertebrates and vertebrates. Microbial populations are able to remove carbon, phosphorus, and human pathogens, as organic materials from leaves, bark, fowl, and human waste are included in the trophic chains supported by microbial consumption. Indigenous and natural microbial populations break down and detoxify most hazardous wastes. Nevertheless, it may become necessary to boost and stimulate them to augment biodecomposing rates and abilities toward pollutants. In addition, in some specific conditions, environmental factors have made it impossible for microorganisms to degrade and detoxify simultaneously large amounts of specific compounds. In water, almost any kind of pollutant can be entirely removed, thanks to the activities of microbial communities. Several case studies demonstrate the feasibility and effectiveness of bioremediation. After a dike break flooded a large part of the Rhine River in 1995, 2500 ha of forest, 1600 ha of swamp, and 1000 ha of grasslands were flooded. The water was diverted to the Rhine River without additional problems. Regulators in many countries have established water quality regulations with maximum levels of certain pollutants. The presence of pollutants at a higher level than that established by law must be a signal to water treatment utilities, which must intervene to remove pollutants or to obtain water from a less polluted river. The natural depuration capacity of water in rivers and wetlands is remarkable. For example, the Parklands were flooded without any significant drought impact. Several guidelines and acts have been developed for the application of bioremediation to the treatment of polluted waters. In our laboratories, continuous and discontinuous systems, including alginate beads, have been created to assess the potential and biological stability of constructed cultures suitable for bioremediation of large amounts of water. An array of important pollutants can be sensibly and completely eliminated from water by such systems [57–59].

14. Air pollution control

Since air pollution has detrimental effects on economies and human health, there is a growing demand for controlling atmospheric pollutants. Various waste air cleaning techniques have been innovated and patented in the last few years. Many of the conventional approaches for dealing with airborne pollution are based on physical and chemical processes. Nowadays, bioremediation strategies are becoming more attractive and environmentally sound. A significant amount of potentially toxic and recalcitrant compounds can be degraded in air treatment systems by microbial cells or enzymes. Up to date, many well-documented studies have addressed developments in the biodegradation of low-volatile organic pollutants and greenhouse gases, which are extensively encountered, including microbial and enzymatic blooms and the biochemistry and microbiology of pollutant degradation. Moreover, advances achieved in studying a multitude of enzyme properties as well as their fundamental applications have been highlighted [37, 60–62]. In nature, airborne volatile organic compounds are oxidatively degraded by a diverse range of organisms, including plants, fungi, and bacteria, as part of their carbon cycling mechanisms. These organisms can be cultivated and used in various biofiltration and biotrickling filtration systems for the removal of air from diverse sources such as paint exhausts, cleaning processes, facilities, car garages, gasoline tankers, or assemblages that release organic solvents; emissions from dry cleaners, refineries, and fuel-propelled automobiles; meat production; purification of liquid, slurry, and solid waste from processing paper and raw rubber; protection from fumigation and herbicides; drinking water treatment to eliminate an undesirable mold aroma; agencies for the treatment of polluted lands, army storage, biofilm release, and shared sewage bacteria for their defense. It should be underlined that the utility of microbial cultures in air pollution control does not mean that different microbial properties must be analyzed. Atmospheric circumstances may adversely affect and have negative consequences on microbial activity in the emergency department. It is important to understand the relationship between functional use and thermodynamic and biological principles. Some studies have investigated the efficiency of the system with specific case presentations. Considering that other pollution control devices may be used as a study conifer model, regulatory measures and security standards must also be complied with. Despite their industrial application to air treatment, new technology papers highlight the importance of sampling microbiological matter from the procedure. Other treatment strategies may be coupled with the policies and technologies to achieve integrated air pollution control [63, 64].

15. Challenges and future directions

Despite having enormous potential, the application of bioremediation is currently not widespread owing to various technological, regulatory, and scientific challenges. In some cases, extensive heterogeneity of pollution along with the multiple metabolic steps required for complete mineralization limits the efficiency of the bioremediation approach. The main limitation of bioremediation is that a higher biodegradation rate is not an exclusive characteristic of microbes existing in soil. Chemical or physical limitations and added costs of bioremediation practices must be considered. Public acceptance is another limit to bioremediation. Encouragement of community support and their participation in the development and finalization of the remediation study are crucial to producing a positive public perception of the application of a bioremediation strategy [1, 38, 65]. Numerous research teams and industries have made use of molecular biology to help the performance of existing microorganisms in biodegrading hazardous pollutants or enhancing their sensing abilities. Among the main applications of the genetic engineering of microorganisms, niche creation and survival ability of engineered strains in a given biotope, as well as the improvement of pollutant-degrading ability, are some of the main applications. The process of creating genetically engineered organisms entails the introduction of one or more pollutant-degrading genes. Due to the complexity of the structure of chemical pollutants, the use of two or more pollutants is needed to better bolster the biodegradation process. Genetic engineering has also made it easier to employ the combined applications of two pollutant-degrading pathways. Therefore, the study of environmental processes should be a national research priority since confronting many environmental problems will influence the technologies we select to build our future [3, 66–68].

16. Current limitations

Bioremediation is not suitable for all contaminated sites, as the efficacy of the technology is subject to a range of scientific and practical limitations. Site-specific factors, such as pollutant type and extent of contaminant bioavailability, the heterogeneity of the contaminated site itself, and the complexity of the microbial community that will be engaged to perform biodegradative activities, present significant design and application challenges. The validity of using pure-culture degradative enzymes over microorganisms as mediators and the likely concentration and optimal application approach required for their potential use are not yet understood. Our understanding of abiotic and biotic interactions is often superficial, and biodegradative interactions are hundreds to thousands of direct and indirect sequences, further amplifying complexity in our quest to understand it. Limitations in knowledge can lead to an underdosing of cell or enzymatic products on site. It is also not always known which enzymatic reaction steps are rate-limiting for application or how to cause these to become the most favored metabolic route [69–71]. Regulatory barriers and market factors have ensured that commercial interest in bioremediation has, in the past, matured to the stage of bench-based engineering tests before these initial investigations have been successful on refinery plastics waste contaminated sites. The development of bioremediation technologies can be costly; consultation with expert or specialist advisors is often required. Ineffective site categorization, assessment, and testing may lead to the development of unnecessary consultant failures. Technologies will have to compete with existing and often more mature treatment options, leading to a greater need to develop affordable solutions. Bioremediation technologies can also take longer than other treatment options. As liability often exists around the release of contaminants into the environment, many potential site owners may be reluctant to treat where the longer-term results are unproven. Regulatory approvals may also be necessary and could be obtained from a range of diverse regulatory authorities with differing and conflicting technology approvals. Site assessments should be comprehensive for the success of any bioremediation plan. Poor upfront design will be further exemplified with case study reports of site assessments, where thorough site assessments would have required the determination of biological activity *in situ* and what biotechnologies had remained after the bulking transfer of polluted surface soils on site. It is often not known when bioremediation projects may take longer than 3 years, as partners, such as monitoring agencies or judge administrators, appointed in legal cases, can change. Regular communication may be beneficial if delays in a project are anticipated. Monitorable target compound libraries are not currently available to assess the performance and stability of altered microbial consortia, where these consortia may be transferred within a complex molecular matrix background. With remediation technologies going forward to their commercialization, a key area in which little R&D activity is reported is the development of suitable upper-end monitoring technologies and method validation studies, as the low-end monitoring field is now generally better understood down to low levels [70–76].

17. Emerging technologies

Molecular biology and biotechnology tools are being used to develop microbial strains showing superior degradation capabilities to promote the remediation process, such as gene cloning, genetic screening, and recombination techniques. Biofilms and methods to enhance their attachment and initiate bacterial growth are being further developed. Adhesion strategies will foster the creation of highly complex and effective bacterial communities that interact with the support and optimize the potential for removing and degrading pollutants. These techniques can be highly effective in multispecies biofilm systems, for example, in cleaning up main columns contaminated with toxic chemicals. Molecular biology tools can be used to determine whether bacteria with added degradation capabilities are functioning within the biofilm, and thus can be used to monitor the success of the immobilized biofilm reactor [38, 77, 78]. Nanotechnology detection methods are expected to improve at a rapid pace with the employment of new advances in chemical systems and electronic plasma technologies. Remediation of pollutants in real time often capitalizes on *in situ* sensor systems, which offer a rapid and versatile technique for monitoring or controlling certain key aspects of bioremediation. Bioinformatics is used to manage, integrate, visualize, and analyze the complex interactions among humans, other organisms, and the environment. Important tools relevant to bioremediation include the metagenomics approach, which uses a new sequencing strategy and a set of bioinformatics tools to understand which organisms are present in an environmental sample. An automated bioremediation system is needed to monitor and control the processes of bioremediation to ensure that appropriate nutrients are added at the appropriate time and safe conditions are maintained. Automated real-time systems in bioremediation are now in the final stages of development and testing. These systems are intended as a feature of *in situ* bioremediation for contaminated groundwater that includes toxicity reduction and biodegradation. The toxicity levels can be monitored from the well under treatment on a real-time basis. These systems could also be used in conjunction with reactive zone treatment. Reactive zone treatment is based on the establishment of reaction zones for the biodegradation of contaminants as they are transferred through the groundwater system [7, 44]. Microorganisms and enzymes play a crucial role in bioremediation processes, facilitating the breakdown of pollutants and contributing to environmental restoration.

When introducing emerging technologies, an integrated system approach could also be considered as single processes may not prove to be the best system. Currently, no commercial biological systems are available for adjusting heavy metal concentration in groundwater. Although they were beyond the scope of the current report, it is noted that identified emerging technologies for bioremediation showed some promise for treating *in situ* groundwater. The use of bioremediation in gas-propelled pump and treat systems would also be an area where future research and development is warranted. Processing potential for bioremediation and the interception of low levels of contaminants *in situ* that exist with this technology option may provide greater water processing capabilities. Moreover, these systems are operated continuously to manage the concentration of discharge. Flow-based systems degrade the feed with a constant hydraulic retention time, irrespective of the nature of the feed. In contrast, contact-based systems degrade the feed with a constant feed concentration, which corresponds to a varying hydraulic retention time. Flow-based bioreactors, or mixed and suspended-growth systems, are commonly used in *ex situ* bioremediation applications, although attention is moving toward the use of immobilized-cell contact bioreactors [79–82].

18. Conclusion

Bioremediation proves to be a sustainable and efficient method for removing toxic compounds from nature, offering an eco-friendly alternative to traditional physical and chemical remediation approaches. By harnessing the metabolic activities of bacteria, fungi, yeast, and plants, bioremediation can effectively degrade and transform a variety of pollutants into harmless products, thereby cleaning the environment from hazardous substances. This method can be carried out *in situ* or ex situ, utilizing specific microorganisms tailored to individual pollutants to achieve successful remediation of soils and aquatic systems. Bioremediation plays a crucial role in addressing environmental sustainability and managing a wide range of environmental pollutants, offering a safe and effective solution to combat the challenges posed by pollution. By promoting the use of biological agents for remediation, such as microbes and plants, bioremediation not only enhances human health but also contributes to improving global air quality and enriching soil and water resources. As a rapidly expanding discipline, bioremediation aims to address environmental issues in a responsible manner and aligns with multiple sustainable development goals. This chapter has provided an overview of the importance of bioremediation and its potential to make significant contributions toward achieving a more sustainable and healthier environment.

Author details

Wafaa M. Abd El-Rahim* and Hassan Moawad
National Research Centre (NRC), Egypt

*Address all correspondence to: wafaa10m@hotmail.com

References

[1] Dinakarkumar Y, Gnanasekaran R, Reddy GK, Vasu V, Balamurugan P, Murali G. Fungal bioremediation: An overview of the mechanisms, applications and future perspectives. Environmental Chemistry and Ecotoxicology. 2024;**6**:293-302

[2] Mokrani S, Houali K, Yadav KK, Arabi AIA, Eltayeb LB, AwjanAlreshidi M, et al. Bioremediation techniques for soil organic pollution: Mechanisms, microorganisms, and technologies—A comprehensive review. Ecological Engineering. 2024;**207**:107338

[3] Rafeeq H, Afsheen N, Rafique S, Arshad A, Intisar M, Hussain A, et al. Genetically engineered microorganisms for environmental remediation. Chemosphere. 2023;**310**:136751

[4] Finley SD, Broadbelt LJ, Hatzimanikatis V. In silico feasibility of novel biodegradation pathways for 1,2,4-trichlorobenzene. BMC Systems Biology. 2010;**4**:1-4

[5] Brar A, Kumar M, Vivekanand V, Pareek N. Photoautotrophic microorganisms and bioremediation of industrial effluents: Current status and future prospects. 3 Biotech. 2017;**7**:1-8

[6] Ain Shuhada Ab Razak N, Habib S, Yunus Abd Shukor M, Aisyah Alias S, et al. Isolation and Characterisation of Polypropylene Microplastic-Utilising Bacterium from the Antarctic Soil. arXiv preprint arXiv:2401.02096. 2024

[7] Kuppan N, Padman M, Mahadeva M, Srinivasan S, Devarajan R. A comprehensive review of sustainable bioremediation techniques: Eco friendly solutions for waste and pollution management. Waste Management Bulletin. 2024

[8] Patel AK, Singhania RR, Albarico FPJB, Pandey A, Chen CW, Dong CD. Organic wastes bioremediation and its changing prospects. Science of the Total Environment. 2022;**824**:153889

[9] Wang M, Chen S, Jia X, Chen L. Concept and types of bioremediation. In: Handbook of Bioremediation. Academic Press; 2021. pp. 3-8

[10] Curiel-Alegre S, Velasco-Arroyo B, Rumbo C, Khan AHA, Tamayo-Ramos JA, Rad C, et al. Evaluation of biostimulation, bioaugmentation, and organic amendments application on the bioremediation of recalcitrant hydrocarbons of soil. Chemosphere. 2022;**307**:135638

[11] Lopes PRM, Cruz VH, De Menezes AB, Gadanhoto BP, Moreira BRDA, Mendes CR, et al. Microbial bioremediation of pesticides in agricultural soils: An integrative review on natural attenuation, bioaugmentation and biostimulation. Reviews in Environmental Science and Bio/Technology. 2022;**21**(4):851-876

[12] Muter O. Current trends in bioaugmentation tools for bioremediation: A critical review of advances and knowledge gaps. Microorganisms. 2023;**11**(3):710

[13] Ayilara MS, Babalola OO. Bioremediation of environmental wastes: The role of microorganisms. Frontiers in Agronomy. 2023;**30**(5):1183691

[14] Fouad FA, Youssef DG, Shahat FM, Abd El-Ghany MN. Role

of microorganisms in biodegradation of pollutants. In: Handbook of Biodegradable Materials. Cham: Springer International Publishing; 2022. pp. 1-40

[15] Yin S, Zhang X, Yin H, Zhang X. Current knowledge on molecular mechanisms of microorganism-mediated bioremediation for arsenic contamination: A review. Microbiological Research. Jan 2022;**258**:126990

[16] Kulikova NA, Perminova IV. Interactions between humic substances and microorganisms and their implications for nature-like bioremediation technologies. Molecules. May 2021;**26**(9):2706

[17] Mydeen AKM, Agnihotri N, Bahadur R, Lytand W, Kumar N, Hazarika S. Microbial maestros: Unraveling the crucial role of microbes in shaping the environment. Acta Biology Forum. 2023;**2**(2):23-28

[18] Amacker N, Gao Z, Hu J, Jousset al, Kowalchuk GA, Geisen S. Protist feeding patterns and growth rate are related to their predatory impacts on soil bacterial communities. FEMS Microbiology Ecology. 2022;**98**(6):fiac057

[19] Kapadia C, Patel N, Rana A, Vaidya H, Alfarraj S, Ansari MJ, et al. Evaluation of plant growth-promoting and salinity ameliorating potential of halophilic bacteria isolated from saline soil. Frontiers in Plant Science. 2022;**13**:946217

[20] Zhang W, Li J, Zhang Y, Wu X, Zhou Z, Huang Y, et al. Characterization of a novel glyphosate-degrading bacterial species, *Chryseobacterium* sp. Y16C, and evaluation of its effects on microbial communities in glyphosate-contaminated soil. Journal of Hazardous Materials. 2022;**432**:128689

[21] Espinosa-Ortiz EJ, Rene ER, Gerlach R. Potential use of fungal-bacterial co-cultures for the removal of organic pollutants. Critical Reviews in Biotechnology. 2022;**42**(3):361-383

[22] Ibrar M, Khan S, Hasan F, Yang X. Biosurfactants and chemotaxis interplay in microbial consortium-based hydrocarbons degradation. Environmental Science and Pollution Research. 2022;**29**(17):24391-24410

[23] Mishra S, Huang Y, Li J, Wu X, Zhou Z, Lei Q, et al. Biofilm-mediated bioremediation is a powerful tool for the removal of environmental pollutants. Chemosphere. 2022;**294**:133609

[24] Zhou Q, Li R, Li T, Zhou R, Hou Z, Zhang X. Interactions among microorganisms functionally active for electron transfer and pollutant degradation in natural environments. Eco-Environment & Health. 2023;**2**(1):3-15

[25] El Moukhtari F, Martín-Pozo L, Zafra-Gómez A. Strategies based on the use of microorganisms for the elimination of pollutants with endocrine-disrupting activity in the environment. Journal of Environmental Chemical Engineering. 2023;**11**(1):109268

[26] Miglani R, Parveen N, Kumar A, Ansari MA, Khanna S, Rawat G, et al. Degradation of xenobiotic pollutants: An environmentally sustainable approach. Metabolites. 2022;**12**(9):818

[27] Sharma P, Kumar S. Bioremediation of heavy metals from industrial effluents by endophytes and their metabolic activity: Recent advances. Bioresource Technology. Jan 2021;**339**:125589

[28] Kieliszek M, Pobiega K, Piwowarek K, Kot AM. Characteristics of the proteolytic enzymes produced

by lactic acid bacteria. Molecules. 2021;**261858**:(7)

[29] Sharma H, Neelam DK. Understanding challenges associated with plastic and bacterial approach toward plastic degradation. Journal of Basic Microbiology. 2023;**63**(3-4):292-307

[30] Xu B, Xu S, Cai J, Sun W, Mu D, Wu X, et al. Analysis of the microbial community and the metabolic profile in medium-temperature *Daqu* after inoculation with *Bacillus licheniformis* and *Bacillus velezensis*. Lebensmittel-Wissenschaft & Technologie. 2022;**160**:113214

[31] Ashkan Z, Hemmati R, Homaei A, Dinari A, Jamlidoost M, Tashakor A. Immobilization of enzymes on nanoinorganic support materials: An update. International Journal of Biological Macromolecules. 2021;**168**:708-721

[32] El-Gendi H, Saleh AK, Badierah R, Redwan EM, El-Maradny YA, El-Fakharany EM. A comprehensive insight into fungal enzymes: Structure, classification, and their role in mankind's challenges. Journal of Fungi. 2021;**8**(1):23

[33] Huang C, Jiang X, Shen X, Hu J, Tang W, Wu X, et al. Lignin-enzyme interaction: A roadblock for efficient enzymatic hydrolysis of lignocellulosics. Renewable and Sustainable Energy Reviews. 2022;**154**:111822

[34] Intasian P, Prakinee K, Phintha A, Trisrivirat D, Weeranoppanant N, Wongnate T, et al. Enzymes, in vivo biocatalysis, and metabolic engineering for enabling a circular economy and sustainability. Chemical Reviews. 2021;**121**(17):10367-10451

[35] Sellami K, Couvert A, Nasrallah N, Maachi R, Abouseoud M, Amrane A. Peroxidase enzymes as green catalysts for bioremediation and biotechnological applications: A review. Science of the Total Environment. 2022;**806**:150500

[36] Bilal M, Noreen S, Zdarta J, Mulla SI, Lou WY, Iqbal HM. Enzyme-oriented strategies to mitigate polluting agents from environment. In: Microbial Rejuvenation of Polluted Environment. Vol. 2. 2021. pp. 267-290

[37] Kumari S, Das S. Bacterial enzymatic degradation of recalcitrant organic pollutants: Catabolic pathways and genetic regulations. Environmental Science and Pollution Research. 2023;**30**(33):79676-705

[38] Narayanan M, Ali SS, El-Sheekh M. A comprehensive review on the potential of microbial enzymes in multipollutant bioremediation: Mechanisms, challenges, and future prospects. Journal of Environmental Management. 2023;**334**:117532

[39] Behbudi G, Yousefi K, Sadeghipour Y. Microbial enzymes based technologies for bioremediation of pollutions. Journal of Environmental Treatment Techniques. 2021;**9**(2):463-469

[40] Bhandari S, Poudel DK, Marahatha R, Dawadi S, Khadayat K, Phuyal S, et al. Microbial enzymes used in bioremediation. Journal of Chemistry. 2021;**2021**(1):8849512

[41] Abdel-Mageed HM, Barakat AZ, Bassuiny RI, Elsayed AM, Salah HA, Abdel-Aty AM, et al. Biotechnology approach using watermelon rind for optimization of α-amylase enzyme production from Trichoderma virens using response surface methodology under solid-state fermentation. Folia Microbiologica. 2022;**1**:1-12

[42] Naveed M, Tianying H, Wang F, Yin X, Chan MWH, Ullah A, et al. Isolation of lysozyme producing *Bacillus subtilis* Strains, identification of the new strain *Bacillus subtilis* BSN314 with the highest enzyme production capacity and optimization of culture conditions for maximum lysozyme production. Current Research in Biotechnology. 2022;**4**:290-301

[43] Tao A, Feng X, Sheng Y, Song Z. Optimization of the artemisia polysaccharide fermentation process by *Aspergillus niger*. Frontiers in Nutrition. 2022;**9**:842766

[44] Wang X, Wang W, Yang N, Wang X, Wang F, Wei X, et al. Design and application of an intelligent monitoring and early warning system for bioremediation of coking contaminated sites. In: Eighth International Conference on Electronic Technology and Information Science (ICETIS 2023). Vol. 12715. Dalian, China: SPIE; 2023. pp. 506-514

[45] Abo-Alkasem MI, Hassan NMH, Abo Elsoud MM. Microbial bioremediation as a tool for the removal of heavy metals. Bulletin of the National Research Centre. 2023;**47**(1):31

[46] Chugh M, Kumar L, Shah MP, Bharadvaja N. Algal Bioremediation of heavy metals: An insight into removal mechanisms, recovery of by-products, challenges, and future opportunities. Energy Nexus. 2022;**7**:100129

[47] Sreedevi PR, Suresh K, Jiang G. Bacterial bioremediation of heavy metals in wastewater: A review of processes and applications. Journal of Water Process Engineering. Jan 2022;**48**:102884

[48] Cisneros-Aguirre J, Afonso-Correa M. Integrated application of innovative technologies for oil spill remediation in Gran Tarajal Harbor: A scientific approach. Waste. 2024;**2**(4)414-450

[49] Menéndez-Serra M, Cáliz J, Triadó-Margarit X, Alonso D, Casamayor EO. Selective pressure influences inter-biome dispersal in the assembly of saline microbial communities. Environmental Microbiology. 2024;**26**(12):e70019

[50] Vonnahme TR, Klausen L, Bank RM, Michellod D, Lavik G, Dietrich U, et al. Light and freshwater discharge drive the biogeochemistry and microbial ecology in a sub-Arctic fjord over the polar night. Frontiers in Marine Science. 2022;**9**:915192

[51] Zięba G, Moryl M, Drzewiecka D, Przybylski M, Pyrzanowski K, Grabowska J. Effects of domestic pollution on European brook lamprey ammocoetes in a lowland river: Insights from microbiological analysis. Water. 2024;**16**(16):2349

[52] Xiang L, Harindintwali JD, Wang F, Redmile-Gordon M, Chang SX, Fu Y, et al. Integrating biochar, bacteria, and plants for sustainable remediation of soils contaminated with organic pollutants. Environmental Science & Technology. 2022;**56**(23):16546-16566

[53] Anekwe IMS, Isa Y. Application of biostimulation and bioventing system as bioremediation strategy for the treatment of crude oil contaminated soils. Soil and Water Research. Jan 2024;**19**

[54] Dehnavi SM, Ebrahimipour G. Comparative remediation rate of biostimulation, bioaugmentation, and phytoremediation in hydrocarbon contaminants. International Journal of Environmental Science and Technology. 2022;**19**(11):11561-11586

[55] Omenna EC, Omage K, Ezaka E, Azeke MA. Bio-augmentation and bio-stimulation with kenaf core enhanced bacterial enzyme activities during bio-degradation of petroleum hydrocarbon in polluted soil. Scientific Reports. 2024;**14**(1):8

[56] Singh B, Christina E. Indigenous microorganisms as an effective tool for in situ bioremediation. In: Relationship between Microbes and the Environment for Sustainable Ecosystem Services. Vol. 2. Elsevier; 2022. pp. 273-295

[57] Gonçalves CR, da Silva Delabona P. Strategies for bioremediation of pesticides: Challenges and perspectives of the Brazilian scenario for global application—A review. Environmental Advances. 2022

[58] King RB, Sheldon JK, Long GM. Practical Environmental Bioremediation: The Field Guide. CRC Press; 2023

[59] Pinheiro LRS, Gradíssimo DG, Xavier LP, Santos AV. Degradation of azo dyes: Bacterial potential for bioremediation. Sustainability. 2022;**14**(3):1510

[60] Balasundaram G, Banu R, Varjani S, Kazmi AA, Tyagi VK. Recalcitrant compounds formation, their toxicity, and mitigation: Key issues in biomass pretreatment and anaerobic digestion. Chemosphere. 2022;**291**:132930

[61] Bhardwaj P, Kaur N, Selvaraj M, Ghramh HA, Al-Shehri BM, Singh G, et al. Laccase-assisted degradation of emerging recalcitrant compounds—A review. Bioresource Technology. 2022;**364**:128031

[62] Wang W, Chang JS, Show KY, Lee DJ. Anaerobic recalcitrance in wastewater treatment: A review. Bioresource Technology. Jan 2022;**363**:127920

[63] Jain R, Gaur A, Suravajhala R, Chauhan U, Pant M, Tripathi V, et al. Microplastic pollution: Understanding microbial degradation and strategies for pollutant reduction. Science of the Total Environment. 2023;**905**:167098

[64] Qiu Y, Zhou S, Zhang C, Zhou Y, Qin W. Soil microplastic characteristics and the effects on soil properties and biota: A systematic review and meta-analysis. Environmental Pollution. 15 2022;**313**:120183

[65] Kour D, Khan SS, Kour H, Kaur T, Devi R, Rai PK, et al. Microbe-mediated bioremediation: Current research and future challenges. Journal Of Applied Biology & Biotechnology. 2022;**10**(2):6-24

[66] Mondal S, Bar S, Roy D, Hazra S, Mal J, Bilal M, et al. Genetic engineering strategies and degradation of pollutants using genetically engineered microorganisms (GEMs). In: Biotechnology for Environmental Protection. Springer Nature Singapore: Singapore; 2022. pp. 209-232

[67] Vickram AS, Shofia SI, Palanivelu J, Karishma S, Saravanan A, Yaashikaa PR. A comprehensive analysis and exploration of the recent developments in the utilization of genetically modified microorganisms for the remediation of hazardous dye pollutants. Groundwater for Sustainable Development. 2024;**13**:101315

[68] Yaashikaa PR, Devi MK, Kumar PS. Engineering microbes for enhancing the degradation of environmental pollutants: A detailed review on synthetic biology. Environmental Research. Jan 2022;**214**:113868

[69] Davis GB. Reviewing the bioremediation of contaminants in groundwater: Investigations over 40

years provide insights into What's achievable. Frontiers in Bioscience-Elite. 2023;**15**(3):16

[70] Yap HS, Zakaria NN, Zulkharnain A, Sabri S, Gomez-Fuentes C, Ahmad SA. Bibliometric analysis of hydrocarbon bioremediation in cold regions and a review on enhanced soil bioremediation. Biology. 2021a;**10**(5):354

[71] Yap JBH, Goay PL, Woon YB, Skitmore M. Revisiting critical delay factors for construction: Analysing projects in Malaysia. Alexandria Engineering Journal. 2021b;**60**(1):1717-29b

[72] Arantes A, Ferreira LMDF. Development of delay mitigation measures in construction projects: A combined interpretative structural modeling and MICMAC analysis approach. Production Planning & Control. 2024;**35**(10):1164-79

[73] Gamil Y, Abd Rahman I. Studying the relationship between causes and effects of poor communication in construction projects using PLS-SEM approach. Journal of Facilities Management. 2023;**21**(1):102-48

[74] Latilo A, Uzougbo NS, Ugwu MC, Oduro P. Role and effectiveness of advance payment guarantees in construction contracts. World Journal of Advanced Science and Technology. 2024;**6**(01):088-102

[75] Leong J, May Yee K, Baitsegi O, Palanisamy L, Ramasamy RK. Hybrid project management between traditional software development lifecycle and agile based product development for future sustainability. Sustainability. 2023;**15**(2):1121

[76] Pamidimukkala A, Kermanshachi S. Impact of Covid-19 on field and office workforce in construction industry. Project Leadership and Society. 2021

[77] Baskaran D, Byun HS. Current trend of polycyclic aromatic hydrocarbon bioremediation: Mechanism, artificial mixed microbial strategy, machine learning, ground application, cost and policy implications. Chemical Engineering Journal. 2024;**30**:155334

[78] Bhatt P, Gangola S, Bhandari G, Zhang W, Maithani D, Mishra S, et al. New insights into the degradation of synthetic pollutants in contaminated environments. Chemosphere. 2021;**268**:128827

[79] Blanco A, Sampedro MA, Sanz B, Llama MJ, Serra JL. Immobilization of non-viable cyanobacteria and their use for heavy metal adsorption from water. In: Environmental Biotechnology and Cleaner Bioprocesses. CRC Press; 2023. pp. 135-153

[80] Fulke AB, Ratanpal S, Sonker S. Understanding heavy metal toxicity: Implications on human health, marine ecosystems and bioremediation strategies. Marine Pollution Bulletin. Jan 2024;**206**:116707

[81] Priya AK, Muruganandam M, Ali SS, Kornaros M. Clean-up of heavy metals from contaminated soil by phytoremediation: A multidisciplinary and eco-friendly approach. Toxics. 2023;**11**(5):422

[82] Megharaj M, Naidu R. Soil and brownfield bioremediation. Microbial Biotechnology. 2017;**10**(5):1244-9